PRAISE FOR *MICRO:BIT FOR MAD SCIENTISTS*

"Packed with projects and experiments for young, mad scientists . . . an excellent resource to take curious kids on a journey into physical computing."

—ANDREW GREGORY, HACKSPACE

"Simon Monk has been teaching folks how to learn to code and make for many years, so he knows exactly the kinds of projects that any beginner can make, and all the trouble spots to avoid. . . . It's a STEAM adventure for students of all backgrounds!"

—LIMOR FRIED, ADAFRUIT

"Fun for all Ages. Great book with very fun experiments. . . . Colorful pictures and instructions will help you create many interesting projects and the rest is up to your imagination."

—MARCO, AMAZON REVIEWER

MICRO:BIT FOR MAD SCIENTISTS

30 CLEVER CODING AND ELECTRONICS PROJECTS FOR KIDS

BY SIMON MONK

no starch press®

San Francisco

Printed in China

Second printing

27 26 25 24 23   2 3 4 5 6

ISBN-10: 1-59327-974-4
ISBN-13: 978-1-59327-974-5

Publisher: William Pollock
Production Editor: Janelle Ludowise
Cover Illustration: Josh Ellingson
Illustrator: Miran Lipovača
Interior Design: Beth Middleworth

Developmental Editor: Liz Chadwick
Technical Reviewer: David Whale
Copyeditor: Paula Fleming
Compositor: Happenstance Type-O-Rama
Proofreader: Abby Franklin

For information on distribution, bulk sales, corporate sales, or translations, please contact No Starch Press® directly at info@nostarch.com or:

No Starch Press, Inc.
245 8th Street, San Francisco, CA 94103
phone: 1.415.863.9900
www.nostarch.com

Library of Congress Cataloging-in-Publication Data

```
Names: Monk, Simon, author.
Title: Micro:bit for mad scientists : 30 clever coding and electronics projects for kids / Simon Monk.
Other titles: Microbit for mad scientists
Description: San Francisco : No Starch Press, Inc., [2019] | Includes index.
Identifiers: LCCN 2019015785 (print) | LCCN 2019021782 (ebook) | ISBN
    9781593279752 (epub) | ISBN 1593279752 (epub) | ISBN 9781593279745 (print)
    | ISBN 1593279744 (print)
Subjects: LCSH: Micro:bit--Juvenile literature. | Single-board
    computers--Juvenile literature. | Electronics--Data processing--Juvenile
    literature. | Python (Computer program language)--Juvenile literature. |
    JavaScript (Computer program language)--Juvenile literature.
Classification: LCC QA76.8.M47 (ebook) | LCC QA76.8.M47 M66 2019 (print) |
    DDC 004.16--dc23
LC record available at https://lccn.loc.gov/2019015785
```

This book is dedicated to Gerard Paris.
A companion to my mother and a friend
and inspiration to all that know him.

ABOUT THE AUTHOR

Simon Monk has a degree in Cybernetics and Computer Science and a PhD in Software Engineering. After spending many years in software and co-founding the mobile software company Momote, he now divides his time between writing books about electronics and programming and helping to run Monk Makes (*https://www.monkmakes.com/*), a business he started with his wife Linda, where he designs electronic kits and accessories for the BBC micro:bit and Raspberry Pi.

You can follow Simon on Twitter (@simonmonk2) and find out more about his books at *https://www.simonmonk.org/*.

ABOUT THE
TECHNICAL REVIEWER

David Whale is an embedded software engineer and a STEM Ambassador who volunteers in schools in the UK. He has been an active member of both the Raspberry Pi and micro:bit communities since their inception. He contributed to the original BBC micro:bit project, advising The IET and BBC and helping to write and deliver resources and training to teachers around the country. He wrote a highly successful children's coding book, *Adventures in Minecraft* (Wiley), and edits a wide range of tech books and magazine articles for well-known authors. David is on a mission to inspire the next generation of engineers and scientists—our future will soon be in their hands.

BRIEF CONTENTS

CONTENTS IN DETAIL

4
MAGICAL MAGNETISM 85

5
AMAZING ACCELERATION 105

ACKNOWLEDGMENTS

I am very grateful to David Whale for finding the time to carry out the technical review of this book. It was a pleasure to work with him. I'm also very grateful to the help and support of the Micro:bit Foundation and to the micro:bit community, who have helped out more than once on technical issues.

I'd also like to thank Liz, Janelle, Bill, and everyone at No Starch Press and of course the very talented Miran Lipovača for the wonderful and amusing illustrations.

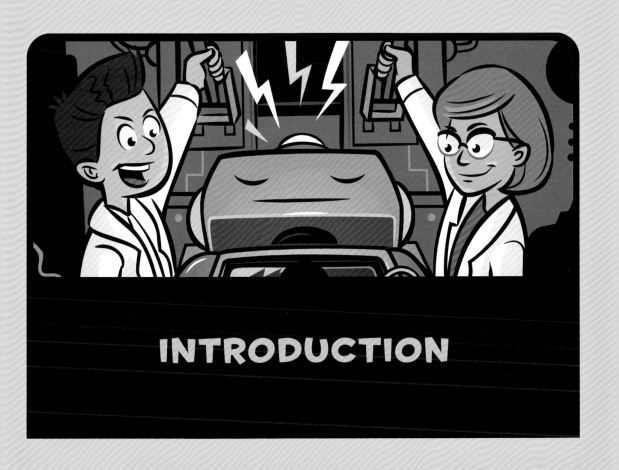

INTRODUCTION

Since the release of the BBC micro:bit in 2016, millions of these devices have been distributed. They're enjoyed by both kids and adults all over the world. The micro:bit was designed as an easy way to teach kids programming skills. One of its great advantages is that you don't need anything more than a USB cable and a computer to start using it. Also, once programmed, it can be disconnected from a power source and run on batteries.

The micro:bit has a small LED display as well as sensors for light, movement, and magnetic fields, so it has everything you need to make interesting projects. When you're ready, you can easily connect things like motors, sensors, and loudspeakers using alligator clips—there's no need for soldering. In other words, your micro:bit can be the *brain* for lots of projects and inventions.

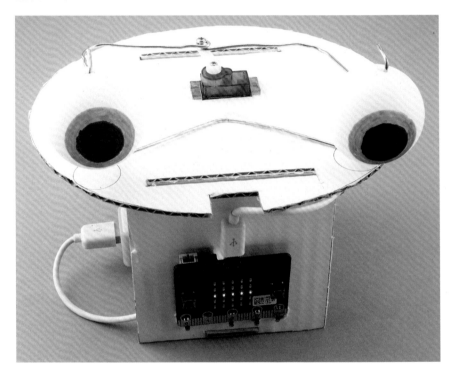

ABOUT THIS BOOK

In these pages, you'll find a variety of experiments and projects. The experiments show you how things work, and then in the projects, you'll use that knowledge to make something funky.

The book is divided into 10 chapters. Chapter 1 tells you all you need to know about connecting and using your micro:bit. You'll use this information in all the experiments and projects in this book. Each of the following chapters deals with a specific topic, such as light, sound, and movement. There are lots of fun and useful things you can do with your micro:bit!

Experiments

Here's a list of the experiments in the book:

Generating sounds   Learn to make your micro:bit play musical notes and other sounds.

It speaks!   Teach your micro:bit to speak!

Sensing light   Use the built-in light detector.

Measuring magnetic fields   Use the built-in magnometer to sense magnets.

Gestures   Use the micro:bit's gesture recognition software to have the device do different things when you shake, drop, or throw it.

Real-time acceleration plotting   Learn about Mu's data visualization feature.

Logging acceleration to a file   Record the data detected by the micro:bit so you can look at it later.

Making a servomotor move   Experiment with motors!

Keeping time   Learn how the micro:bit tells time.

How fast are your nerves?   Test your reaction time.

Measuring temperature   Use the micro:bit temperature sensor to build a thermometer.

Finding the radio range   Learn how to use the micro:bit for radio communication.

Projects

Here's a list of the projects in this book:

Musical doorbell   Plays a tune of your choice when pressed

Shout-o-meter   Detects a sound and shows how loud it is

Automatic night-light   Turns on by itself when the room gets dark

Light guitar   Makes music when you wave your hands over the micro:bit's LEDs

Infinity mirror   Creates the illusion of infinite depth using light

Compass   A real working compass!

Magnetic door alarm   Goes off when you open a door, separating a magnet from the micro:bit

Toothbrushing monitor   Measures your brushstrokes to make sure you're keeping your pearly whites in good order

Acceleration display   A meter that shows how quickly the micro:bit is accelerating

Animatronic head   A robotic head that has moving eyes and a talking mouth

Robot rover   A two-wheeled micro:bit robot!

Binary clock   Tells time using LEDs

Talking clock   Announces the time every hour and whenever you press a button

Lie detector   Measures electrical currents in the skin to tell whether someone is lying

Temperature and light logger   Automatically keeps a record of light and temperature levels

Automatic plant waterer   Waters your plants whenever it senses that the soil is too dry (Never again kill a plant!)

Wireless doorbell   A wireless upgrade to the doorbell project that uses radio waves

Micro:bit-controlled rover   A wireless version of the roving robot that receives your instructions via radio

CODE AND RESOURCES

The two most popular computer languages for programming the micro:bit are Makecode Blocks (referred to as just *Blocks* in this book) and the MicroPython programming language.

Wherever possible, I've provided programs for the projects and experiments in both Blocks and MicroPython. This means you don't have to type in the code yourself—unless you want to.

You can find all the code for the activities in this book at the companion GitHub page: *https://github.com/simonmonk/mbms/*. I provide full instructions for accessing and using the code in Chapter 1.

1

GETTING STARTED

This chapter will get you started with your BBC micro:bit. It will also set the stage for the experiments and projects you'll find in the following chapters. I'll give our Mad Scientist (that's you!) some ideas about what to do with your micro:bit, and I'll help you start programming your micro:bit. You'll learn how to use Blocks code and MicroPython.

Mad Scientists are generally too busy and distracted to type in huge amounts of code, so all the code used in this book is available to download. This chapter explains how to get and use the code.

A TOUR OF THE MICRO:BIT

Let's take a quick look at the micro:bit and what's on it.

The Top

Figure 1-1 shows the top of a micro:bit.

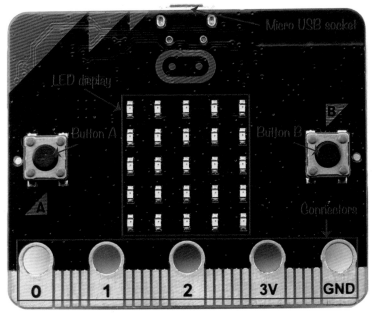

Figure 1-1: The micro:bit

At the top of the board is a micro USB socket, which you'll use to connect the micro:bit to a computer. You'll write programs for your micro:bit on a computer, so you'll have to transfer the programs to the micro:bit using a USB connection. The USB socket can also power up the micro:bit.

On the left and right are two buttons labeled A and B, respectively. We can program these buttons to trigger some action, like flashing an image on the LEDs or making a doorbell sound.

Between the two buttons is a grid of 25 LEDs arranged as 5 rows and 5 columns. This is the micro:bit's display. Even though it has only 25 LEDs, this display can show scrolling text messages, small images, patterns—a whole load of things!

At the bottom edge of the micro:bit is a gold-plated (yes, real gold!) strip called the *edge connector*. This strip has five holes labeled 0, 1, 2, 3V, and GND. These large connectors allow you to connect things to the micro:bit using alligator clips. For example, you might connect a speaker to get the micro:bit to make sound or a motor to get it to move. The much smaller connectors, the slits between the holes, can be used only with a special adapter. In this book, we'll use the large connectors for everything except the two robot projects, which require an adapter to connect to a motor controller.

The Bottom

Now let's turn the micro:bit over and see what we have on the underside (Figure 1-2).

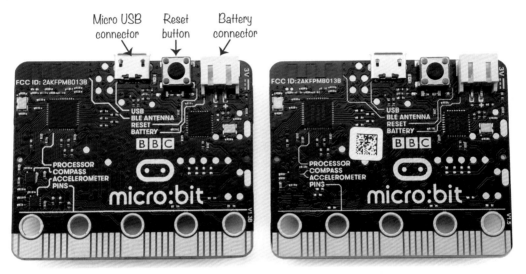

Figure 1-2: The micro:bit's underside, version 1.3B (left) and 1.5 (right)

At the time of writing, there are two versions of the micro:bit in circulation. Both work in the same way, and it doesn't matter which version you have for this book. The latest version (1.5) just has a slightly simplified design. You can see the difference in the bottom left of the boards shown in Figure 1-2.

The leftmost connector is the micro USB connector. Immediately to the right of that is a push switch. This is the micro:bit's *reset button.* Pressing this button will cause the micro:bit to restart whatever program is installed on it. To the right of the reset button is the battery connector, which allows you to connect a 3V battery pack to your micro:bit.

Now we'll take a closer look at each section, starting with how to power your micro:bit.

Power and the Micro:bit

You can power the micro:bit through the USB connector or with batteries, depending what you want to do with it.

Using USB

The USB connection will provide the micro:bit with 5V (5 volts) of power when you plug it into your computer. However, the micro:bit needs only 3.3V, not 5V, and too much voltage would damage the micro:bit. Therefore, it has a USB interface chip that converts this 5V to the 3.3V that the micro:bit expects.

When the micro:bit is plugged in via USB, you can use the 3V connector on the edge connector to provide power to low-current electronic devices, like external LEDs or speakers designed to work with the micro:bit.

NOTE   *The reason this connection is labeled 3V rather than 3.3V is partly that there isn't much space for another number but also that protection circuitry reduces the 3.3V to a voltage close to 3V.*

The connection labeled GND is the *ground* or 0V power connection. When you power your micro:bit, current flows out of the 3V connection, and it needs to return to the micro:bit in order to complete the circuit—the GND connection is where the current returns.

Using Batteries

Once you've programmed your micro:bit, you may want to move it away from your computer, in which case you'll need batteries. You can use a AAA battery pack, like the one shown in Figure 1-3. Simply plug the ends of the battery pack wires into the 3V socket on the underside of the micro:bit.

Figure 1-3: Battery packs for the micro:bit

The appendix lists some places you can buy these battery packs. A built-in switch is useful because you can power the micro:bit down without unplugging the battery pack, a task that can be a bit tricky.

Don't use rechargeable AAA batteries because their voltage is generally too low to power a micro:bit. These batteries can also allow dangerously high currents if the flow of electricity is accidentally short-circuited. If you want to use rechargeable batteries, one option is a USB battery backup like the one shown on the left of Figure 1-4.

Figure 1-4: Using rechargeable batteries with the micro:bit

These rechargeable batteries allow you to power your micro:bit through its USB port. The lower-capacity, cheaper backup batteries are often more suitable to power a micro:bit than are the more expensive versions. The micro:bit uses so little current that these more advanced units, which tend to turn off if they think the battery hasn't been used in a while, can be fooled into thinking nothing is connected.

The item on the right of Figure 1-4 is the Monk Makes Charger for micro:bit. This uses a lithium polymer (LiPo) battery that automatically recharges whenever you connect the charger to your computer with a USB cable. When you want to run the micro:bit on batteries, you just unplug the USB cable to the charger.

You can find out more about the options for powering your micro:bit in the appendix.

WARNING *The earliest versions of the micro:bit were susceptible to damage when used with USB batteries and power supplies. If you have any board other than a V1.3B or V1.5, you should avoid using any power supply except your computer USB port or a 3V battery pack.*

The early micro:bits don't have a version number on them. Flip your micro:bit over and look at the bottom right corner on the underside near connector 0. If it says V1.3B or V1.5, then using USB battery packs and power supplies will be fine. If you don't see a version identifier here, don't use these power supplies.

In either case, steer away from power supplies and USB batteries described as high power.

You can read the micro:bit Foundation's full safety advice at https://microbit.org/guide/safety-advice/.

Connecting Electronics with Input/Output Pins

One especially fun thing about the micro:bit is that you can use the connections to attach electronic gadgetry of your own creation. In this book, you'll build projects that use these connectors to control motors, lights, and a loudspeaker. You'll also use them as inputs to gather data from sensors that measure light, sound, and temperature.

The 3V and GND connections are there to supply power. The connections labeled 0, 1, and 2 are usually referred to as *input/output pins*, and you'll be attaching electronics to these.

NOTE *You might wonder why we call them* pins *when they look nothing like pins. The term* pin *comes from the chip on the circuit board that the micro:bit connects to. The chip does have a pin—a tiny leg-like connection—for each of these three pins.*

The pins 0, 1, and 2 can be used for:

▸ Digital output—turning an external LED on and off, for example

▸ Analog output—controlling the brightness of an LED, for example

▸ Pulse generation—to control a servo motor, for example

▸ Digital input—to detect when an external button has been pressed, for example

▸ Analog input—to measure temperature using an analog temperature sensor, for example

▸ Touch input—to detect that you have touched the pin or a wire connected to a pin

Digital and Analog Inputs

When you use a pin as a digital output, you can write commands in your programs to turn the output on (set it to 3V) or turn it off (set it to 0V), but *only* on or off—nothing in between. This is also the case with digital inputs: they can only ever be on or off. When your program code reads a digital input, if the voltage at the input is nearer 3V than 0V, then the input counts as being on; otherwise, it's off.

Analog inputs, however, can be somewhere between on and off. Analog inputs on a micro:bit can be any value between 0 and 1,023, depending on the voltage at the pin.

Making an Analog Signal: Pulse Width Modulation

A micro:bit, like all other electronics, can work only with digital on or off signals. To get analog signals between 0 and 3V, electronic devices *simulate* the analog signal by providing a rapid series of digital pulses. The longer the pulse, the more power is delivered to whatever is connected to the analog output. This is known as *pulse width modulation*, or *PWM*. Figure 1-5 shows PWM in action.

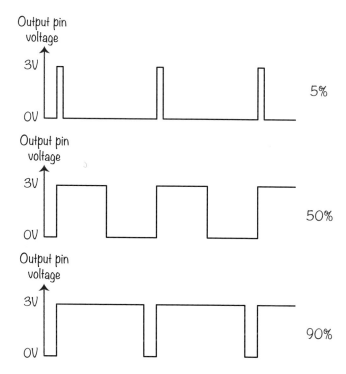

Figure 1-5: Analog outputs on the micro:bit delivering different amounts of power

If the output is connected to an LED and the pulse is at 3V only 5 percent of the time, the LED will appear to glow dimly. In contrast, if the pulse is at 3V for 90 percent of the time, the LED will appear to be at almost full brightness.

Interestingly, in both cases, the LED would actually be flashing at 50 times a second, but the human eye can't see a light flash that quickly. Instead, we just register a relatively dim or bright light.

If the three pins 0, 1, and 2 are not enough for your project, you can use an adapter to access the many pins between the three numbered ones.

Built-In Peripherals

The writing on the underside of the micro:bit gives us some clues about other things the micro:bit can do. Two areas at the bottom left are labeled *compass* and *accelerometer*.

The compass is actually a *magnetometer*, meaning it measures the strength of magnetic fields. You can use it as a compass, but you can also use it to detect the presence of magnets.

The accelerometer measures forces acting on the accelerometer chip. Because gravity is a constant force pulling down on everything, you can, by measuring the forces acting in different directions, use the accelerometer to detect when the micro:bit is being tilted and by how much, as well as when it is being shaken or in free fall.

Also on the underside, you will see the words *BLE Antenna*. The micro:bit includes BLE (Bluetooth Low Energy) hardware that allows your micro:bit to communicate wirelessly with other micro:bits or Bluetooth-enabled phones.

Note that the technology for micro:bit-to-micro:bit communication is not actually Bluetooth; it just uses the same frequency. Chapter 10 is devoted to this feature of the micro:bit.

HARDWARE ESSENTIALS

There are a few things you'll need for nearly all the experiments and projects in this book. They include:

Micro:bit

Micro USB cable   To connect your micro:bit to a computer (Make sure this is a standard micro USB data cable and not a *charging cable*, which lacks the necessary connections for communication. You can't use a charging cable to load programs onto the micro:bit.)

Alligator clip cables   Ideally, these should be no longer than 4 to 5 inches to avoid getting tangled up.

3V AAA battery pack with two AAA batteries

USB power supply   You'll need this only for some projects. (See earlier warning.)

Each project or experiment will have a list of all the items you need, and the appendix at the end of the book gives more information about how to acquire these supplies.

This book tries to keep project construction straight-forward and, apart from the roving robot in Chapter 6, no soldering is required. For most projects, you just need alligator clip cables to connect things together. When making connections using alligator clips, it's best to clip the cable in vertically so the teeth of the clip look like Figure 1-6, as this makes it much less likely that the cables will come loose.

Figure 1-6: Securely attaching alligator clips

PROGRAMMING THE MICRO:BIT

Mad Scientists aren't known for their patience, so let's make our micro:bit do something. First, we'll program our micro:bit.

One of the nice things about a micro:bit is that to get started with it, you just need a USB cable and a computer with a browser and internet connection. You can use a computer running on Windows, macOS, or Linux. As long as your device has a modern browser, such as Chrome, it will work fine.

We'll first connect up the micro:bit. Then we'll make a small program using two methods: the drag-and-drop Blocks code and MicroPython, which you need to type out.

Connecting your Micro:bit

Start by connecting your micro:bit to your computer with a micro USB cable. Nearly all micro USB cables will work fine, but remember that charging-only cables don't have the necessary data connections and won't work. If you have trouble programming your micro:bit using the following instructions, try using a different USB cable.

Once you connect your micro:bit, your operating system should react as if you'd just plugged in a USB flash drive. To transfer a program onto the micro:bit, find the micro:bit in your filesystem just like you'd find a flash drive or some other plug-in. Then copy a file called a *hex file* into the micro:bit folder icon, and, hey presto!, your program will be installed. Loading a program onto your micro:bit is also known as *flashing*.

Let's make a hex file and flash it to our micro:bit.

Programming with Blocks: Hello World

You can build programs for the micro:bit through the micro:bit website without downloading any software. We'll make a program that scrolls some text across the LED display on the micro:bit. Open your browser and navigate to *https://makecode.microbit.org*, and you should see a window something like Figure 1-7.

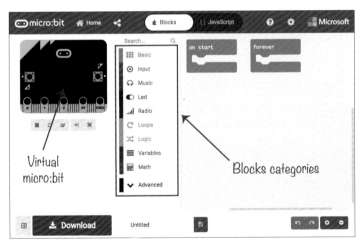

Figure 1-7: The https://makecode.microbit.org *web page*

This is the editor that you build programs in. On the left is an image of a micro:bit that works like a virtual micro:bit, running whatever programs you write.

The section in the middle is a list of categories such as Basic, Input, and Music. Within each of these categories, you'll find *blocks* that you can drag onto the working area on the right. Each block is an instruction for your micro:bit. By dragging these blocks around and connecting them, you'll write code using the Blocks language.

You'll notice that when you open the editor, there are already two blocks in the editing section: on start and forever. Any blocks inside the on start block will run once when the micro:bit first powers up, when a new program is uploaded, or when the micro:bit resets because you pressed the reset button. Whatever blocks are inside the forever block will run over and over again, until you stop the program.

For our first program, we don't need the forever block, so select it and press DELETE to remove it from your program. Next, you need to add a show string block to your program—*string* is programming speak for *text*. To do this, click the **Basic** category, drag the show string block into the programming area, and place it inside the on start block, as shown in Figure 1-8. If you have sound on your computer, you will hear a satisfying *click* as the blocks snap together.

Now click inside the text bubble and type **Hello World**. You can also type different text—whatever you want to see displayed.

Figure 1-8: Blocks code for displaying Hello World

As soon as you drag the show string block into place, the virtual micro:bit on the left of the display should scroll your message across its display to show you what your program does.

Now let's transfer your program to the real micro:bit. Connect your micro:bit using a USB cable and click **Download** at the bottom left of the web page.

This will download the file from the editor in the same way as any other file that you might download from the internet. Where the file is saved will depend on your operating system and browser, but usually it's in a folder called *Downloads*. Find this folder, click into it, and you should find a file called *microbit.hex*. Using the File Manager (or Finder on a Mac) select this file and drag it onto the micro:bit where it appears in your filesystem (Figure 1-9).

Figure 1-9: Dragging a file onto your micro:bil

As soon as you release your mouse button, the file should start installing itself onto the micro:bit. You'll know this is happening because the LED on the back of the micro:bit will start blinking. When the blinking ends, the micro:bit will reset itself and run the program, scrolling the text across its display. If you miss seeing the message, press the reset button on the back of the micro:bit to see it again.

DOWNLOADING DIRECTLY TO YOUR MICRO:BIT

Most browsers have an option to choose where a file is saved each time you download one. You can use this feature to download files directly onto your micro:bit. Then you don't have to first download and then copy the file.

To set this up in the Chrome browser, go to *chrome:// settings/*, click **Advanced**, scroll down to the Downloads, and enable the option **Ask where to save each file before downloading**. That way, next time you click Download in the editor, you'll be prompted for a location to save your file to and can select the micro:bit folder as the destination.

At the time of writing, flashing programs onto your micro:bit is about to get a whole lot easier for users of the Chrome browser. You can read about this feature here: *https://support.microbit.org/support/solutions/ articles/19000084059-beta-testing-web-usb.*

Adding Graphics

To display our message, we added a show string block to the on start block. The on start block is a special type of block called an *event block* that runs the code connected to it whenever a particular event happens. In this case, the event is the start of the program.

Let's make our program a bit more complicated by adding a new event that will detect when button A is pressed. To do this, click the **Input** category and drag an on button A pressed block into the program area. Next, drag a show leds block from the Basic category into the on button A pressed block. The squares in the show leds block represent each LED in the LED display. You can select which LEDs should light up by clicking them so they turn white. The result should look something like Figure 1-10.

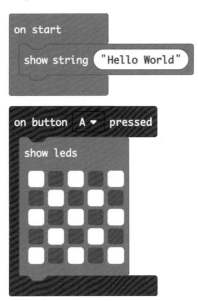

Figure 1-10: Adding an LED pattern to the program

The show icon block has a number of ready-made LED images if you want to choose one of those. Click **Download** again and copy the new hex file onto your micro:bit. Once the new program has been uploaded, you can test it by pressing button A. When you do so, the selected LEDs in the show leds block should light up on your micro:bit, as in Figure 1-11.

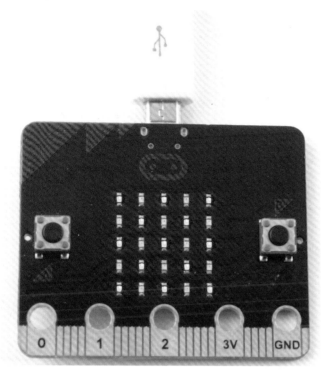

Figure 1-11: Displaying an LED pattern on an actual micro:bit

Saving and Sharing

The *https://makecode.microbit.org* website remembers all your projects. Give your project a name by entering the name in the Save area next to the Download button, and your project will be saved. Whenever you change your program, it should automatically save, but to make sure, you can save manually by clicking the floppy-disk Save icon. Note that when you click Save, the hex file will also be downloaded, but you can just click Cancel if you're not ready to flash the program onto your micro:bit.

Note that your stored programs are actually kept in your browser cache, so if you delete your cache, you will lose your programs.

To switch between programs and start new ones, click **Home** at the top of the page.

To publish a program so others can see it, click **Share** next to Projects. This will ask you to confirm that you wish to publish your project. Click **Share** again, and you'll see a link like the one in Figure 1-12.

Figure 1-12: Sharing a project

You can give that URL to anyone, and when they put it in their browser, they'll get a copy of your program to use. Note that because this is just a copy of your program, there's no risk that they'll spoil your original copy of the code.

Finding Blocks

The show string and show leds blocks we've used so far were pretty easy to find, but the Blocks editor contains a lot of blocks, and it keeps the ones you don't use as often hidden. This way, you don't see a confusing array of choices when you start out on your micro:bit adventure.

These blocks are hidden in two ways. First, you may have noticed that when you select one of the block categories, such as Basic, an item labeled . . . *more* appears (see Figure 1-13).

Figure 1-13: Extra blocks in the more section

In the case of the Basic category, the more section contains blocks such as clear screen and show arrow. If you click through the other categories, you'll see that they all have a more section, some with quite a few extra blocks.

Incidentally, hovering your mouse over a block will open a little information box telling you what the block does. Investing a little time in getting familiar with the various blocks will give you exciting ideas about what you can to do with your micro:bit.

The second place to find hidden blocks is in the Advanced category, just after Math (Figure 1-14). Selecting this category will reveal a host of other categories containing more advanced features of the Blocks language, starting with Functions. Again, spend a little time familiarizing yourself with the blocks. Don't worry if it's not obvious what some of them do. If you're interested in a block that does a particular thing, you may find the search field useful for finding the block you want.

Figure 1-14: The Advanced *blocks category*

The categories in the Advanced section that you'll use frequently are Text and Pins. In some chapters, you'll also be using Functions and Arrays.

Programming with MicroPython: Hello World

Blocks code is great for getting started with programming, because you can accomplish some really impressive things with just a few blocks. However, many people prefer writing a program in text rather than dragging blocks around. Typing lines of text is also more like regular programming.

MicroPython is an implementation of the Python 3 programming language that includes everything you need to program your micro:bit with Python. In fact, many of the blocks in Blocks code have direct equivalents in MicroPython, so switching from programming in Blocks to MicroPython is fairly straightforward. Python is a popular first programming language and is often taught in schools for that reason.

Downloading the MicroPython Editor

We'll use the *Mu* editor app (*https://codewith.mu*), which has lots of features. It can be downloaded straight onto your computer. Mu also lets you flash your program directly to a micro:bit without having to drag the hex file around. The nice thing about Mu is that you don't need internet access to flash the program once Mu has been downloaded.

Download Mu from *https://codewith.mu/#download*. There are several versions of Mu, so it's important that you get the right one for your computer. The download page will offer you different versions for different operating systems. If you're a Windows user, download the version under Windows Installer that's labeled *64-bit* (Figure 1-15). If you're a macOS user, there's just one version.

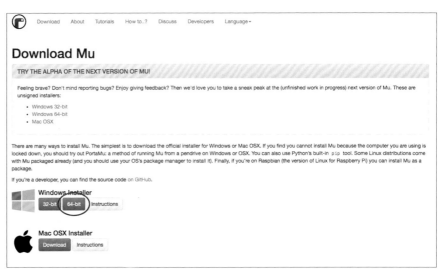

Figure 1-15: Downloading Mu

Run the installer and accept the license agreement and all the default options.

The first time you run Mu, it will offer you a choice of *modes*. Make sure to select the BBC micro:bit mode (Figure 1-16).

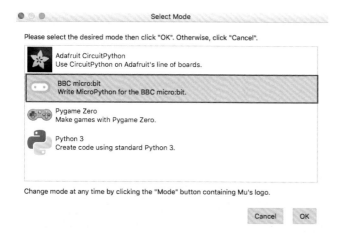

Figure 1-16: Selecting the right version of Mu

Mu will present you with a blank editor window where you'll type your first program.

Writing the Program

Let's give Mu a go! Add the following code in the Mu window:

```
from microbit import *

display.scroll("Hello World")
```

It should look something like Figure 1-17.

Figure 1-17: Writing a MicroPython program in Mu

Save the program by clicking **Save** at the top of the Mu window. You'll be prompted to enter a name for your program: call it *hello.py*.

Now, with your micro:bit connected to your computer, click **Flash** at the top of the Mu window. This should start the process of loading the hex file to the micro:bit, just as though you were dragging over a file produced by the Blocks editor. Once the flashing is complete, the *Hello World* message should scroll across the display.

Let's take a closer look at the code. Here's the first line:

```
from microbit import *
```

You'll find this line at the start of pretty much every Micro-Python micro:bit program, because this is what tells MicroPython to include all the built-in code that makes it compatible with the micro:bit's display and other hardware. This code isn't automatically included because MicroPython can be used on a lot of different boards, not just the micro:bit. The line actually means: *from the microbit library, import everything* (* means everything). You could also enter `import microbit`, but then you'd have to prefix everything with `microbit`, which is a lot of extra typing.

The only other line of code in our minimal program displays a message on the display:

```
display.scroll("Hello World")
```

This line uses the command `display.scroll`, which tells the micro:bit to scroll something across its display. Then you add a set of parentheses, inside of which we have some text enclosed in quotation marks. You use parentheses to add extra information to a command. In this case, the extra information (also called an *argument*) is the text you want to display. You also have to enclose the text in double quotes to show that the program should treat it as text, not as more programming commands.

If, when you flash the program onto a micro:bit, you see something other than *Hello World* (or nothing at all), you probably have an error in your code. When using a text-based programming language, you have to be precise in what you type. For example, misspelling a word such as `display` or `scroll` will cause an error when you run the program. Errors in programming are called *bugs.* Bugs only show up when the program tries to run on the micro:bit. If you get any bugs, you can carefully compare the code you wrote to the code in the book to make sure it all matches. But there's another way to find and fix bugs—you can use the REPL.

The REPL

The REPL (Read-Eval-Print-Loop) is a *command line interface* to MicroPython on the micro:bit—that is, it's an area that lets you send Python commands directly to the micro:bit, without having to put them into a program first. If you entered the line 1 + 2 here, the REPL would respond with 3. Similarly, if MicroPython encounters a problem when it tries to run, it can report this problem to you in the REPL without you having to wait for the micro:bit to try to display something.

To experiment with the REPL, let's deliberately introduce an error into our program by misspelling `scroll` so that it only has one l. Delete an l and click **Flash** again. The program will upload despite being faulty, but this time, the micro:bit display will say: `AttributeError: 'MicroBitDisplay' object has no attribute 'scrol'`. It takes a long time to scroll out this message, so to get a clearer view of it, click the **REPL** button at the top of the Mu window and then click the reset button on the back of your micro:bit. You should then see the error message in full in the REPL (bottom of Figure 1-18).

Figure 1-18: Catching errors using the REPL in Mu

The >>> prompt in the REPL indicates that you can type Python commands here to run immediately on the micro:bit. Try typing the following after the >>> prompt and hitting ENTER at the end of the line (this time spell scroll correctly):

```
display.scroll("The REPL is useful")
```

Your micro:bit's display should immediately scroll out the message, without your needing to upload a program. You don't need to repeat the import command, because MicroPython has already executed import—it did so before it encountered the misspelled scroll.

NOTE   *When you were typing in your Hello World program, you may have noticed Mu trying to help by guessing what you were typing. For example, if you type display and then pause, a list of options (clear, get_pixel, is_on, off, on, scroll, set_pixel, and show) will appear. You can click on the one you want to save yourself some typing.*

Adding Graphics

Let's now add some graphics to the program, just like we did with the Blocks code. We again need to detect when button A has been pressed and then display the graphics pattern. This is a little trickier in MicroPython because this language doesn't have the same concept of events that you find in the Blocks code. Instead, we have to write a *loop* that repeats the commands contained within it until told to stop. In our program, these commands will check for a button press and, if that event has happened, perform the necessary action. In other words, rather than being told that button A has been pressed, the program has to keep checking whether it's been pressed. Here is the code:

```python
from microbit import *

display.scroll("Hello World")

while True:
    if button_a.was_pressed():
        display.show(Image.CHESSBOARD)
```

The `while True` line of code marks the start of the loop that will continue until something stops it, like you unplugging your micro:bit or pressing the reset button or CTRL-C at the REPL. Whenever you make a loop, make sure to indent any lines of code that should run in that loop. Luckily, Mu recognizes when you have started a loop and helpfully indents the next line for you.

The first line within the loop is an `if` statement. This uses the `button_a.was_pressed` function to check whether the button has been pressed since last time `was_pressed` was used. If it has, then the lines indented below the `if` statement will be run. You'll notice that the next line is indented even more, which means this line should only run if the `if` line is true (so if the button was indeed pressed). In this case, this line of code tells the display to show a ready-made graphic that belongs to the Image library. I've chosen the `CHESSBOARD` image. We'll talk about `if` commands more later in "if Blocks" on page 29.

In Python, indentation is very important, and remembering to indent can be a big cause of frustration when first learning the language. Lines that are indented within, for example, a `while` or an `if` command must be indented by exactly the same amount. In Mu, this is four spaces. As you practice coding in Python, knowing when to indent will become easier and easier.

PROGRAMMING CONCEPTS

Here we'll go over some key ideas in programming that are worth understanding, particularly when you want to modify some project code or start making your own projects. These ideas are the same whether you're using the Blocks editor or MicroPython, so we'll cover how to execute each concept both ways, first in Blocks code and then in the MicroPython equivalent.

Variables

A *variable* is a name associated with a value or multiple values. When you want to use the value, you can call the name in your code. Let's illustrate this idea with some examples.

The Blocks Code

One of the main categories in the Blocks editor is Variables. Start a new project by clicking **New Project** from the Blocks editor's home page. Delete the forever block (we don't need it) and then from the Variable category, click **Make a Variable....** When prompted for a new variable name, enter counter. Drag the set counter to 0 block into the programming area and click it onto the on start block.

Next, from the Input category, add an on button A pressed block. Inside this block, add a change counter by 1 block from the Variables category and then a show number block from the Basic category. Finally, from the Variables category, drag out a counter block and click it into the 0 in the show number block to replace the 0.

Once you've done this programming, your code should look like the following.

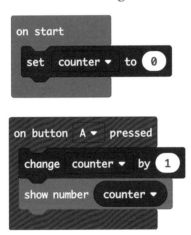

You can try out this program in the virtual micro:bit on the left of the editor by clicking button A. You should see the number displayed increase by 1.

Let's have a look at what's going on here. In the on start block, you make a variable called counter and give it an initial value of 0. When the on button A pressed block is activated, the change block changes the value of the counter variable by adding 1 and then shows the value of the counter variable on the display.

In this case, our counter variable contains a number, but we can also set variables to hold text and even collections of data.

The MicroPython Code

The MicroPython version of the program we just wrote looks like this:

```
from microbit import *

counter = 0

while True:
    if button_a.was_pressed():
        counter += 1
        display.scroll(str(counter))
```

We import the usual micro:bit library, and then make a counter variable and give it an initial value of 0. We make a while loop that ensures that if button A is pressed, 1 is added to counter. To add 1, we use +=, which is equivalent to the change counter block.

When we want to display the new value, we have to convert the numeric value into a text string with str(counter), because the display command works only with strings.

Note that it's perfectly okay to use one command inside another. So here, we use the str command inside the display .scroll command by putting it inside parentheses. When you do this, the innermost command (str in this case) is run first and supplies a value to the next command (in this case display.scroll).

Arithmetic

In the previous example, you saw how we could add a number to a variable. As well as addition, we can use all the usual arithmetic options in programs, including subtraction, multiplication, and division.

The Blocks Code

Say we wanted to keep doubling a number instead of counting by ones. We could alter the previous program so that the starting number for counter is 1 and the on button A pressed block multiplies counter by 2, as shown here.

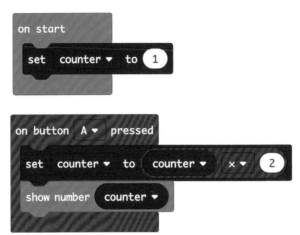

Now, instead of using a change counter by 1 block, we use a set counter to block and, inside that, use a multiply block (×). In the first part of the multiply block, we place the variable counter, and in the second part, we put the number 2. Now every time button A is pressed, counter is given the value of 2 times its old value.

When we use blocks, math like this rapidly becomes quite complicated, as we need to nest blocks inside of blocks inside of other blocks. If we have a program that needs to do a lot of arithmetic, it may be better to code the project using MicroPython.

The MicroPython Code

In MicroPython, we use arithmetic symbols such as +, -, * (multiply by), and / (divide). You can also use brackets to change the order of the math operations, as you do in math class. Here's how we would rewrite the doubling Blocks program we just made:

```
from microbit import *

counter = 1

while True:
    if button_a.was_pressed():
        counter = counter * 2
        display.scroll(str(counter))
```

The key line here is counter = counter * 2. The = symbol after the variable name means that whatever follows the equal sign will be assigned to the variable as its new value. In this case, that's counter * 2 (counter times 2). You could also use the shorthand form counter *= 2, just as we did earlier when doing addition.

if Blocks

Programs can be thought of as a series of steps that the micro:bit will carry out. Sometimes you'll need your program to make decisions and execute particular steps based on those decisions. The blocks in the Logic category allow your programs to make decisions.

We'll make a variation of our counting-by-1 example that counts up to 10 and then goes back to 0 to start the counting process over.

The Blocks Code

Try making the following Blocks program.

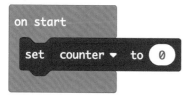

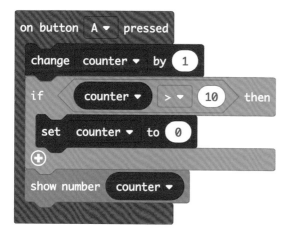

First, we add an if block after the change counter by 1 block. Onto the if block, we drop a comparison block that compares the value of counter to 10. If the counter value is greater than (>) 10, the program runs the blocks inside the if block. In this case, the if block contains a single block that sets counter back to 0. If counter is not greater than 10, then it continues to show the number on the display.

There are other variants of the if block that let you carry out one action if a condition is true and another action if the condition is false. You'll learn more about these in later chapters.

The MicroPython Code

We've already used if in MicroPython when we checked whether a button had been pressed. Here, we're not just using it to see whether a condition is true or false. Instead, we're comparing the variable counter to the value 10. The MicroPython version of the Blocks code looks like this:

```
from microbit import *

counter = 0

while True:
    if button_a.was_pressed():
        counter += 1
        if counter > 10:
            counter = 0
        display.scroll(str(counter))
```

Take a good look at the indentation of this program: we have an if inside another if, which is itself inside a while. See if you can figure out how this code works, using the explanation of the Blocks version as a guide.

Strings

Remember that a string is a series of characters (numbers, letters, or punctuation), often a word. You used a string in your very first Hello World program in Figure 1-8 to display the message *Hello World*.

The Blocks Code

Most of the blocks for strings can be found in the Text category. The simplest of these is a block with a pair of quotes and a gap where you can add your own text to make a string. You can use this block to assign a string, rather than a number, to a variable. Here we set the message variable to a string. This program will display the string's length.

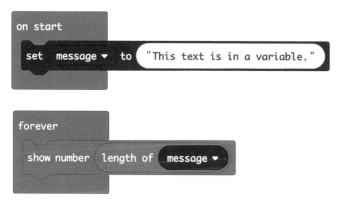

We use an on start block, inside which we set the value of the variable message to some text. In the show number block, you can see a length of block and, inside that, the message variable. The length of block supplies the number of characters in the string message to show number, which then scrolls that number across the display.

The Text category has other blocks that let you do things, like join together two strings, chop out a section of a string, and convert a string into a number.

The MicroPython Code

In MicroPython, string values are distinguished from other program code by being enclosed in double quotes, just as in Blocks code. We would write the previous Blocks program like this in MicroPython:

```
from microbit import *

message = "This text is in a variable."

while True:
    display.scroll(str(len(message)))
```

This works in the same way as the Blocks code. Notice that we use the str command to convert the length of the string len into a string itself so that it can be displayed.

Arrays and Lists

So far, we've used variables that contain just a single piece of data, whether that's a number or string. But sometimes you need to keep a collection of values in a variable, often to access each value in turn, like a sequence of notes that make up a tune (as you'll see in Chapter 2).

In Blocks code, a collection of values is called an *array*, and in MicroPython, it's called a *list*. The idea is the same for both, however.

The Blocks Code

The Blocks editor has an Array category that contains array-related blocks. The most important are set list to, set text list to, and get value at. The first two allow you to create a list containing numbers or strings, respectively, and get value at lets you access a particular element of an array.

The following small program creates a variable called colors and assigns it to an array of four strings. Then it picks a random item from that array to display. Note that the array of block has + and - buttons that allow you to adjust the number of items in the array when you create it.

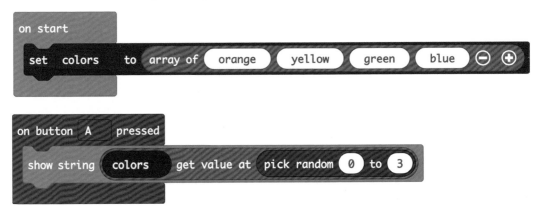

We add an on `button A pressed` block. Inside it, we add a `show string` block, inside which we add a `get value at` block. We then add `pick random` (found in the Math category) to choose any number from 0 to 3. Altogether, this means that when button A is pressed, the `pick random` block selects a random number between 0 and 3 and fetches the array element at that position, and then the `show string` block displays that element. Note that in both Blocks and MicroPython programming, array numbering starts at 0 rather than 1, so if the number 0 is randomly chosen, the *first* element of the array will be displayed.

The MicroPython Code

This is how you would write the previous Blocks program in MicroPython:

```
from microbit import *
import random

numbers = ["orange", "yellow", "green", "blue"]

while True:
    if button_a.was_pressed():
        display.scroll(numbers[random.randint(0, 3)])
```

In MicroPython, we use square brackets, [and], to enclose the elements of a list, and we separate the list elements with commas. We also use square brackets to access a particular element of a list using its position. In this case, the `random.randint` command, which returns a number between 0 and 3, is enclosed within the square brackets.

Programming Wrap-Up

This has been a very quick introduction to programming the micro:bit. All the code for this book is available for download, so you don't have to master programming before you start doing experiments and making projects. As you progress through this book, you'll be introduced to new blocks and MicroPython language features, which I'll explain as they come up.

For more information on MicroPython for the micro:bit, see *https://microbit-micropython.readthedocs.io*. If you're new to Python, you might find my book *Programming micro:bit: Getting Started with MicroPython* (McGraw-Hill, 2018) a useful accompaniment to this book.

DOWNLOADING THE CODE

Some of the programs in this book are long and complex, and typing them out might not be something that a Mad Scientist such as yourself is eager to do. If you don't want to make the programs yourself, you can just download them and flash them to your micro:bit.

Downloading the Blocks Code

All the Blocks code is published on GitHub at *https://github.com/simonmonk/mbms/*. When you click a link for the Blocks code, the project will open in your browser.

Scroll down the GitHub page until you will see something like Figure 1-19. You're looking for a list of links to all the Blocks code programs.

Figure 1-19: Links to the Blocks code for all the programs in this book

To open one of the programs, just click the link. For example, Figure 1-20 shows the result of clicking the link for the musical doorbell project.

Figure 1-20: Following a link to the Blocks code for the musical doorbell project

You'll notice that Figure 1-20 does not look like the normal Blocks editor window. That's because you're to just view the code here and flash it onto your micro:bit to use. If you want to edit the code, or just see the code in its more familiar editor, click the **Edit** button at the top right. This will make a copy of the original program for you, and then you'll be able to edit this copy.

Downloading the MicroPython Code

You can download the MicroPython programs from *https://github.com/simonmonk/mbms/*.

If you're used to using the Git software, you can clone the entire project onto your computer. For non-Git-experts, here's a step-by-step guide for getting all the code.

1. Visit *https://github.com/simonmonk/mbms/* and then click the green **Clone or download** button and select **Download ZIP**, as shown in Figure 1-21.

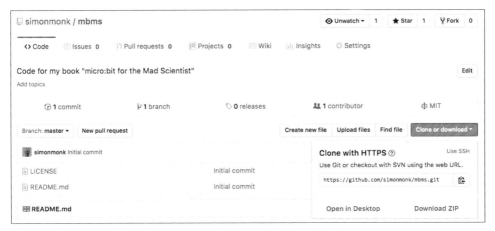

Figure 1-21: Downloading the code for this book

2. Find the ZIP file you just downloaded (*mbms-master.zip*) in your downloads folder and extract all the files from it.

   The process of extracting the files will vary, depending on whether you use Windows, macOS, or Linux. On macOS and most Linux distributions, a ZIP file will automatically extract when you open it. If you're a Windows user, be aware that while Windows will let you look inside the ZIP file without extracting it into separate files, you won't be able to use the files unless you extract them. To extract the files in Windows, right-click the archive file in File Explorer and select the option **Extract All** (Figure 1-22).

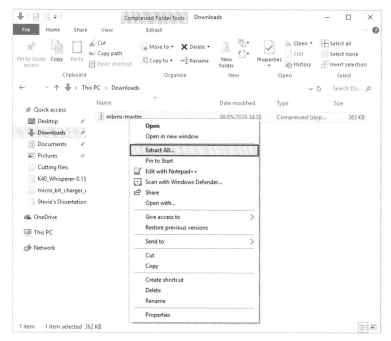

Figure 1-22: Extracting the ZIP archive file in Windows

The extracted files will be contained in a directory called *mbms-master*. Within this directory, you'll find another directory called *python*, and within this, you'll find all the MicroPython programs for the book as separate files, each with the file extension *.py*.

3. Unfortunately, you can't just double-click a program to open it in Mu. To open a program, first open Mu and go to Load. Then find the MicroPython program you want to open. As a shortcut, I suggest you move all the MicroPython programs from the *python* folder that you just downloaded into the folder where Mu normally expects to find its programs. By default, this is a directory within your home directory called *mu_code*. Now when you click the Load button in Mu, you'll see all the MicroPython programs straight away.

SUMMARY

Now that we've looked at some of the basics of the micro:bit, it's time to get started with some proper Mad Scientist experimentation and project work. We'll begin with making and detecting sounds using a micro:bit.

2

SUPER SONIC

n this chapter, we'll explore using sound with the micro:bit. We'll teach the micro:bit to play music and even imitate speech, and we'll get it to hear sound by connecting it to a microphone. You'll try out a couple of experiments and create two simple projects: the first project is a musical doorbell that lets the Mad Scientist know when visitors have arrived, and the second is a Shout-o-meter that measures and displays the volume of sounds it detects.

CONNECTING A LOUDSPEAKER TO A MICRO:BIT

There are a couple of ways to hear sound from your micro:bit. Which one you should choose depends on how much sound you want to make.

The Quiet Method: Headphones

Perhaps the easiest way to get sound from your micro:bit is to use alligator clip cables to connect the micro:bit to a pair of headphones (see Figure 2-1).

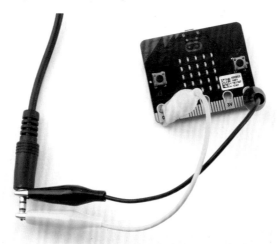

Figure 2-1: Alligator clips attached to headphones

If you look closely at the metal plug on the headphones, you should see that it is really made of three pieces separated by rings of plastic. This means the plug has three connections. The section closest to the headphones' wire is the ground connection. Connect this to the micro:bit's GND (0V) connection with an alligator clip.

The other two connectors are the audio signals for your left and right ears. If you want to hear sound in both ears, place the alligator clip so that it spans both of the two connectors on the end. You can also attach the alligator clip to the very tip for sound in just one ear (as shown in Figure 2-1). Either way, clip the other end of the alligator clip to any of the three micro:bit pins: 0, 1, or 2. Micro:bit users traditionally use pin 0 for audio.

Headphones designed for use with a cellphone that include a microphone will have four connectors on the plug rather than three. This shouldn't make a difference. You can still use the tip as the audio connection and the connector closest to the plug body as the GND connection.

To upgrade this method slightly, you can use an *audio jack adapter* like the one shown in Figure 2-2. Just plug your headphones straight into the adapter, with the black wire connected to GND and the other to pin 0. Adapters like this fit directly onto the headphones and provide a more reliable connection than alligator clips.

Figure 2-2: An adapter to connect an alligator clip to a 3.5 mm audio jack

The Ghetto Blaster Method: Speaker

With an amplified speaker such as the one shown in Figure 2-3, you can produce a lot more sound using the same connection methods described earlier: either connecting directly to the speaker plug or using an audio jack adapter.

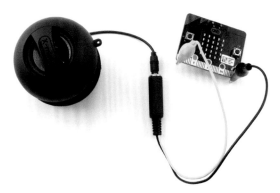

Figure 2-3: Connecting a micro:bit to an amplified speaker

Some speakers are designed especially for use with micro:bits. Some of these have cables that end in alligator clips to attach to your micro:bit, while others, like the Monk Makes Speaker for micro:bit shown in Figure 2-4, end in pins similar to the micro:bit's, making it easy to connect the two with alligator clip cables.

Figure 2-4: The Monk Makes Speaker for micro:bit

Amplified speakers need a power source. In some cases, the amplified speaker may have its own batteries or USB cable. Otherwise, the micro:bit itself could power the speaker, in which case the devices will have to connect in three places: to GND (0V) and 3V on the micro:bit in order to power the speaker and to pin 0 (or one of the other pins) for the audio signal coming from the micro:bit.

Whatever you're using for audio output, let's test it out!

EXPERIMENT 1: GENERATING SOUNDS

In this experiment, you'll learn how to generate sounds using your micro:bit and a loudspeaker or headphones.

What You'll Need

To carry out this experiment, you just need:

Micro:bit

Speaker or headphones

Alligator clip cables

You can find sources for these in the appendix.

Here we'll assume you're using a Monk Makes Speaker for micro:bit and a set of alligator clips, but any of the speaker connection methods listed earlier will work.

Construction

1. Connect the speaker using one of the methods shown in Figures 2-1 to 2-4. Then plug your micro:bit into your computer.

2. Go to *https://github.com/simonmonk/mbms/* to access this book's code repository and click the link for **Experiment 1: Generating Sounds**. Once the program has opened, click **Download** and then copy the hex file onto your micro:bit. If you get stuck, head back to Chapter 1, where we discuss the process of getting programs onto your micro:bit in full.

If you prefer to use Python, download the code from the same website. For instructions for downloading and using the book's examples, see "Downloading the Code" on page 34. The Python file for this experiment is *Experiment_01.py*.

3. Once you've successfully programmed the micro:bit, press **button A**. You should hear a tone through your speaker or headphones!

Code

You won't need much code for this experiment. Whether you use Blocks code or MicroPython, it's just a matter of detecting button A being pressed and then playing a -.

Blocks Code

The Blocks code for this experiment is shown here.

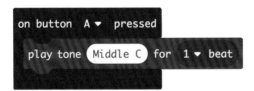

The code uses the on button A pressed block to run the play tone block every time button A is pressed. You drop the play tone block into the on button A pressed block so it clicks into place. Then from the drop-down menu, select the tone you want to hear (in this case Middle C) and the duration of the note (1 beat).

MicroPython Code

Here's the MicroPython version of the code:

```
from microbit import *
import music

while True:
    if button_a.was_pressed():
        music.pitch(262, 1000)
```

Python has a huge number of *libraries*, which are collections of code that do a specific thing. By asking your code to use these libraries, you get access to a lot of functionality without having to write complicated code yourself. The music library is an example: it contains functions you can use to make your micro:bit make sound. To make MicroPython use the music library, you first import the library using the `import music` command.

While Blocks code will handle some things on its own, like knowing how often to run code and what order to run it in, MicroPython requires you to make that clear in the code itself. Here, you use a `while True:` loop to tell the micro:bit to keep checking whether someone has pressed button A.

When someone does press button A, the note plays using the `pitch` command, which needs two pieces of information: the frequency of the note (262 is middle C) and the duration of the note in milliseconds (in this case, 1000 milliseconds or 1 second).

Things to Try

You might like to try changing the tone produced. If you are using Blocks code, go back to the browser and click the **Edit** button to alter the code, then click **Middle C**. This will open up a mini keyboard where you can choose a different note to play. To change the note in MicroPython, enter a new number instead of 262 for the frequency. Then click the **Flash** button again. Later in this chapter, you'll learn a better way to choose notes using MicroPython.

You could also try making both buttons A and B play tones and even have them play different tones—a chord!

How It Works: Frequency and Sound

How does the micro:bit create sound in the speaker? Essentially, the micro:bit switches a current (the flow of electricity) on and off incredibly fast, causing part of the speaker to vibrate, creating sound. The speed at which the micro:bit switches the current on and off determines the *frequency* of the sound, and that's what makes different tones. I'll explain this in more detail.

Figure 2-5 shows the parts of a loudspeaker. A rigid, usually metal frame holds a cone in place. The narrow end of this cone is cylindrical and has a coil of wire wrapped around it. Around this coil, fixed to the frame of the loudspeaker, is a strong magnet.

When a current passes through the coil, it—and hence the whole cone—moves back and forth very rapidly. This vibration creates pressure waves in the air that we hear as sound.

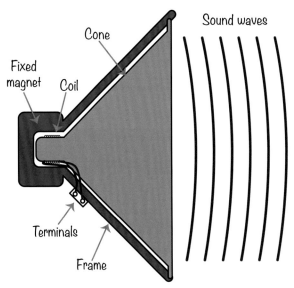

Figure 2-5: A loudspeaker

To make a particular sound, the speaker cone needs to move back and forth a certain number of times per second. The number of times per second the speaker moves is its frequency, measured in *hertz* (shortened to Hz). The higher the frequency, the higher the pitch of the sound. A frequency of 262 Hz corresponds to middle C on a piano. The C an octave higher has a frequency of 524 Hz, or double middle C. In music, when you go up an octave, you double the frequency.

The micro:bit controls the current and therefore the frequency by turning pin 0 on and off very rapidly. When pin 0 is off, it has an output voltage of 0V, and when it is on, it has a voltage of 3V. If you were to draw a chart of the output voltage against time, it would look like Figure 2-6.

For obvious reasons, this type of wave is called a *square wave*. Since a micro:bit's outputs can only ever be on or off, this is the only kind of wave that we can generate from the micro:bit.

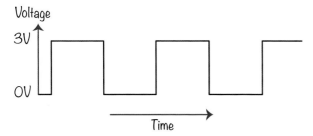

Figure 2-6: A square wave

When the micro:bit sends its signal to the amplified speaker, the speaker takes the low current signal from the micro:bit and increases the current to drive the speaker with more power, making everything louder.

Now let's experiment with making sounds.

EXPERIMENT 2: IT SPEAKS!

The micro:bit's MicroPython software has a really neat feature that allows you to make your micro:bit read out phrases. In this experiment, we will try out this feature and have our micro:bit talk to us.

Although the software that generates the speech was designed for use with English, by experimenting with the spelling, you should be able to make the library speak in other languages.

This feature isn't (at the time of writing) available through the Blocks code, so we'll be using MicroPython.

What You'll Need

This project uses exactly the same hardware as Experiment 1.

Micro:bit
Speaker or headphones
Alligator clip cables

Construction

1. Connect the speaker using one of the methods shown in Figures 2-1 to 2-4.

2. This project uses the speech library, which is not available in Blocks code, so this experiment code is for Python only. Go to *https://github.com/simonmonk/mbms/* and download the *Experiment_02*.py file. You'll also find code for the other projects and instructions for downloading and using the book's examples on the GitHub page. Flash the program onto your micro:bit.

3. Once the micro:bit has been successfully programmed, press **button A** on the micro:bit. You should hear a message being spoken through your speaker or headphones. The Mad Scientist likes to hear this voice as it's a reminder of their dear old friend Professor Hawkins, who alas is no longer with us.

Code

The MicroPython code for the experiment is listed here:

```python
from microbit import *
import speech

while True:
    if button_a.was_pressed():
        speech.say("Mad Scientists love micro bits")
```

Aside from importing the speech library, getting the micro:bit to speak is as simple as putting some text for it to say in the say function.

The speech library is quite sophisticated—you can even use it to vary the pitch to make your micro:bit sing! You can find out all about the library at *https://microbit-micropython .readthedocs.io/en/latest/tutorials/speech.html*.

 # PROJECT: MUSICAL DOORBELL

Difficulty: Easy

The Mad Scientist is particularly partial to a musical doorbell. In fact, you will not be surprised to hear that one of the scientist's favorite tunes is "Imperial March" from *Star Wars*.

In Chapter 10, we will revisit this project, adding a second micro:bit that will make the doorbell work wirelessly.

This project (shown in Figure 2-7) is a variation on Experiment 1, except that instead of playing a single tone when a button is pressed, the doorbell will play tunes. We'll have button A play one tune and button B play another. You can see a short video of the project in action here: *https://youtu.be/ xmLupw4PxYQ/*.

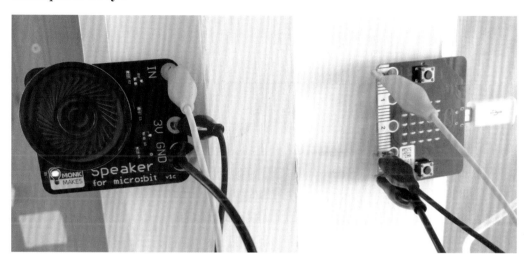

Figure 2-7: The musical doorbell project

Giving the visitor two tunes to choose from allows them to indicate the level of urgency of their visit. Then if the Mad Scientist is busy, they can just ignore the person at the door!

What You'll Need

For this project, you will need the following items:

Micro:bit   To be the controller for this project and provide two buttons to press

3 × Alligator clip cables   To connect the micro:bit to the speaker (Longer cables will make this easier)

USB power adapter or 3V battery pack with power switch   To power the micro:bit and speaker

Speaker   To play the doorbell tune (I recommend the Monk Makes Speaker for micro:bit)

Adhesive putty or self-adhesive pads   To attach the micro:bit to the door frame and the speaker to the inside of the door frame

If you use batteries for this project, it's a good idea to use a battery box with a power switch so that when not in use, the doorbell can be switched off to save the batteries. Otherwise, the batteries will be exhausted after only a day or so of use. A USB power supply offers a longer-term solution that can be left on all the time.

Construction

When building a new project, it's always worth constructing and testing it at your desk before you fit it into place where it will be used.

1. Connect the speaker to the micro:bit using the three alligator cables, as shown in Figure 2-7.

    It's a good idea to use color-coding for your cables, with black for GND, red for 3V, and any other color for the audio connection from pin 0 of the micro:bit. Using different colors will help you keep track of the connections.

2. Go to *https://github.com/simonmonk/mbms/* to access the book's code repository and click the link for **Musical Doorbell**. Once the program has opened, click **Download** and then copy the hex file onto your micro:bit. If you get

stuck on this, head back to Chapter 1, where we discuss the process of getting programs onto your micro:bit in full. If you prefer to use Python, download the code from the same website, along with instructions for downloading and using the book's examples. The Python file for this experiment is *ch_02_Doorbell.py*.

3. Once the micro:bit has been successfully programmed, press **button A** on the micro:bit and you should hear a tune playing (Scott Joplin's "The Entertainer"). Now press **button B** and you will hear Frédéric Chopin's "Funeral March."

4. Once you have everything working, disconnect the micro:bit from your computer and plug it into your USB power adapter or battery box. Test it out again to make sure you've got it working. Then fix the micro:bit part of the project onto one side of your door and the speaker side of the project to the other side of the door. There are a few things to note here:

   Firstly, sticking things to walls, even with adhesive putty, can make a mess, so make sure you get permission if you need to. This is especially true if you are using sticky pads, as these can attach quite permanently to paint.

   Secondly, the alligator clips will need to pass from one side of the door to the other in such a way that they don't get too pinched when the door closes. So work out where they need to go before you start sticking anything down. In Chapter 10, we will make another version of this project that uses a second micro:bit to provide a wireless link.

   Finally, if you are using a USB power adapter, you will need a power outlet that's close enough for the USB power adapter to reach your micro:bit.

Code

Now let's talk through the code for the project.

Blocks Code

Here's the Blocks code.

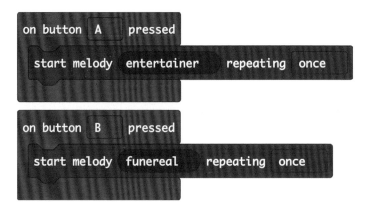

The code is similar to that of Experiment 1, with a few differences. First, we have two stacks of code: one for button A and one for button B. Second, we choose once from the repeating menu, because we want the melody to play only once.

Third, we use the start melody block to play a whole sequence of notes rather than just a single note. Notice that these tunes are already available in the blocks—you just need to select them from the menu!

MicroPython Code

Here is the MicroPython version of the program:

```
from microbit import *
import music

while True:
    if button_a.was_pressed():
        music.play(music.ENTERTAINER)
    elif button_b.was_pressed():
        music.play(music.FUNERAL)
```

This works exactly the same as the Blocks code. The music .play method is equivalent to the start melody block, and we use if statements to check which button was pressed. The if statements allow button A and button B to play different tunes.

The same predefined tunes are available to play in both Blocks and MicroPython code.

Things to Try

Picking from a selection of predefined tunes is all very well, but the Mad Scientist may have particular tastes in music. They may want to compose their own tunes!

If you are using Blocks code, you can make your own tune by creating a list of play tone blocks, like the example shown here. You fill out the notes you want played, and each note is played in turn.

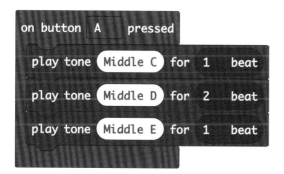

So if you know all the notes for a particular tune, you can create it like this. You'll also need to specify how long each note needs to play. You may have to experiment a bit before you get your music to sound just the way you want.

Now let's see how to create a tune in MicroPython:

```
from microbit import *
import music

notes = ['A4:4', 'A', 'A', 'F:2', 'C5:2', 'A4:4', 'F:2', 'C5:2',
 'A4:4']

while True:
    if button_a.was_pressed():
        music.play(notes)
```

The music library for MicroPython takes care of playing whole tunes by letting you use a special notation to write your own melodies. Each note is made up of a string of characters (see Chapter 1 for more information on strings). The first character of the string is the note name (a letter A to G). Next comes an octave number—middle C is in octave number 4, so you will probably want to restrict your tune to around octaves 3, 4, and 5. The octave number is optional, and if you don't give it, Python will assume you want the first octave.

Once you specify an octave number, the music library will assume that octave applies to all following notes until you specify a different octave number.

Next, you can optionally put a colon followed by a duration. The duration is measured in quarter-notes. For example, to play middle C for a half-note, you would write C4:2.

To string together several notes, you have to create a *list*. So far we've used variables that hold only a single element. A list is like a variable that can hold multiple elements, and you can access and use each element independently. To indicate that the notes variable contains a list of values, rather than just a single value, you'll separate the array values by commas and enclose the whole thing between [and].

In our array, each element is a note string. To play the whole sequence of notes, you use the play function, providing it with the list of notes to play. This example plays the opening few notes from the *Star Wars* "Imperial March."

Here, you see we import the usual microbit library, as well as the music library. We save our tune in a variable called notes. Then we make another while True: loop so that the code keeps running and checking whether the button was pressed. We tell the program that if button A is pressed, it should play the notes variable.

PROJECT: SHOUT-O-METER

Difficulty: Easy

The Mad Scientist likes to measure things. To that end, this project makes a simple sound meter that indicates the volume of a noise. Then the scientist can tell the neighbors off for making too much noise—and prove they really are.

What You'll Need

For this project, you need a microphone to pick up sounds so you can measure their volume. I'm going to use the microphone built into the Monk Makes Sensor, which has a bunch of sensors. The sound's volume then appears on the micro:bit's LED display. The louder the sound, the more LEDs will light up.

For this project, you will need the following items:

Micro:bit   To be the controller for this project and provide two buttons to press

3 × Alligator clip cables   To connect the micro:bit to the speaker (Longer cables will make this easier)

Any micro:bit power source   Can be the USB computer cable or a battery box

Monk Makes Sensor for micro:bit   To supply a microphone

Construction

1. Connect the sensor board to the micro:bit using the three alligator clips, as shown in Figure 2-8. You need to connect 3V on the sensor to 3V on the micro:bit, GND to GND, and the hole with the microphone picture to pin 0 on the micro:bit.

    It's a good idea to stick to the color-coding of the cables, with black for GND, red for 3V, and any other color for the microphone connection from pin 0 of the micro:bit.

Figure 2-8: The Shout-o-meter project

2. Go to *https://github.com/simonmonk/mbms/* to access the book's code repository and click the link for **Shout-O-Meter**. Once the program has opened, click **Download** and then copy the hex file onto your micro:bit. If you get stuck on this, head back to Chapter 1, where we discuss in detail how to get programs onto your micro:bit.

    If you prefer to use Python, then download the code from the same website, along with instructions for downloading and using the book's examples. The Python file for this experiment is *ch_2_Shoutometer.py*.

3. Once you've programmed the micro:bit, try whistling near the microphone (Figure 2-9) and notice how the LEDs jump in response to the sound level. You can also try tapping the microphone. You can see a short video of the project in action here: *https://youtu.be/6pGDSHmfFng/*.

Figure 2-9: The Monk Makes Sensor for micro:bit microphone

Code

The Blocks version of this code is able to make use of the built-in plot bar graph of block, whereas the MicroPython version is more complicated because we have to implement our own version of this feature.

Blocks Code

The Blocks language includes a useful block called plot bar graph of that makes the code for displaying the sound level really easy.

We put a forever block in, so the code inside is constantly running. Then we add the plot bar graph of block, which will display the loudness from the microphone.

As you can see, the analog value read from pin 0 of the micro:bit has 511 subtracted from it before being passed to plot bar graph of with a maximum value of up to set to 512. The reason for this bit of math is discussed in "How It Works: Microphone Output" on page 59.

Getting the right blocks assembled can be tricky, especially when it comes to math. Fortunately, the editor allows you to freely move blocks around, so if they are not in the right place to give you the results you want, you can just drag them to where they should be. See Chapter 1 for more information on editing code.

MicroPython Code

The MicroPython version of the code is a little more complicated than the Blocks code. MicroPython does not have a built-in bar graph display, so we have to write our own. The plot bar graph of block provides a nice, smooth display, despite rapidly changing data. To get the same result in MicroPython, I had to add code to read the maximum sound level from 10 samples.

```
from microbit import *

def sound_level():
    max_level = 0
    for i in range(0, 10):
        sound_level = (pin0.read_analog() - 511) / 100
        if sound_level > max_level:
            max_level = sound_level
    return max_level

def bargraph(a):
    display.clear()
    for y in range(0, 5):
        if a > y:
            for x in range(0, 5):
                display.set_pixel(x, 4-y, 9)

while True:
    bargraph(sound_level())
    sleep(10)
```

We use the sound_level function and make a for loop to
take 10 samples of sound. Each sample value is (as with the
Blocks version of the code) the analog value with 511 subtracted
from it. However in this case, to scale down the number of rows
to be lit to 0 to 4, we divide the resulting value by 100. We then
compare the sound level stored in the variable sound_level to
the variable max_level and, if it is greater, max_level is changed
to be the sound_level. When all 10 samples have been taken,
the largest one will be in max_level, and this value is returned
by the function.

The bargraph function takes a value, represented by a, to
display. The higher the value, the more LEDs will be lit, indicating a louder noise. This value should be between 0 and 4.
However, if it is greater than 4, it doesn't matter—all the LEDs
in the display will turn on, but nothing else will happen. The
function works by looping over each row of the display, and, if
the value of a is greater than the row number, every LED on
that row is illuminated by the inner for loop that asks whether
x is in the range of 0 to 4.

All the main while loop has to do is call the function
bargraph, supplying it with the sound level returned by the
function sound_level.

How It Works: Microphone Output

Figure 2-10 shows a graph of the output of the microphone when it is detecting sound. Voltage is on the vertical axis, and time in on the horizontal axis.

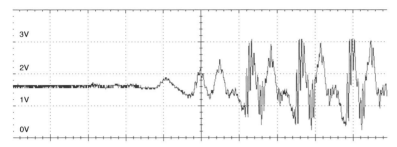

Figure 2-10: A sample of sound

As you can see from the left-hand side of the plot, before the sound starts, the output voltage from the sensor is level at about 1.5V. When the sound starts, the voltage oscillates above and below this 1.5V value as the microphone picks up the pressure waves of the sound. A reading of 1.5V on the micro:bit gives an analog value of 511. This is why we subtract 511 from the reading before displaying it on the micro:bit; otherwise, half the LEDs would be on during silence.

SUMMARY

In this chapter, the Mad Scientist explored the world of sound, both by producing music and speech from the micro:bit and by detecting sound using a microphone. We have started our exploration of the micro:bit with a couple of easy projects.

In the next chapter, we will take a look at light. We'll measure light with a special sensor and use the micro:bit's LED display. Then we'll tackle a large project, using the multicolored NeoPixel display and combining light with sound to make a light-controlled musical instrument. After that, we'll move on to other, even more challenging projects.

3

LUMINOUS LIGHT

n this chapter, we'll use the micro:bit to experiment with light. First, you'll learn how to sense light levels, and we'll build a light-controlled guitar that plays notes depending on how much light it senses. Then we'll use light to create an optical illusion infinity mirror that appears to go on forever to help the Mad Scientist fathom their most profound thoughts. MicroPython doesn't have a light-sensing feature yet, so you'll use Blocks code in this chapter.

EXPERIMENT 3: SENSING LIGHT

Let's look at how you can use a micro:bit to measure the light levels. Once you know how to do this, you'll be able to make all sorts of light-dependent projects, including the automatic night-light and light-controlled guitar you'll find later in this chapter.

The micro:bit's developers very cleverly built a feature into the device's software that allows it to measure the light level with its LEDs. I'll explain how in "How It Works" on page 63.

What You'll Need

You just need a micro:bit and a USB cable for this experiment.

Construction

1. Visit the book's code repository at *https://github.com/ simonmonk/mbms/* and click the link for **Experiment 3: Sensing Light**. Click **Download** and then copy the hex file onto your micro:bit. If you get stuck, head back to "Programming the Micro:bit" on page 11 where we discuss the process of getting programs onto your micro:bit in full. Remember, this experiment only works on Blocks code, so there's no MicroPython code available to download.

2. Once you've successfully programmed the micro:bit, a number between 0 and 9 should appear on the display, indicating how much light the micro:bit is detecting. Try changing the light level by shading the micro:bit with your hand or holding it under a light and see how the number changes.

Code

Here is the Blocks code for this experiment.

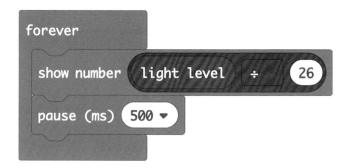

The light level block takes a reading from the micro:bit's light sensor and assigns it a value between 0 and 255 that represents the brightness. But, if the number is more than one digit, the micro:bit's LED display will need to scroll to display it, which is slow and inconvenient. By dividing that number by 26, the value will always be below 10.

After displaying the light level, the micro:bit pauses for half a second to give you time to read the display. Otherwise, the numbers would flicker past too quickly when the light level changes.

How It Works

The micro:bit doesn't have a dedicated light sensor. Instead, it uses an ingenious trick to calculate light levels with its display. You can read about this trick at *https://lancaster-university.github.io/microbit-docs/extras/light-sensing/*.

PROJECT: AUTOMATIC NIGHT-LIGHT

Difficulty: Easy

Although they're reluctant to admit it, the Mad Scientist is a little bit afraid of the dark. That's why they developed a project that uses a micro:bit's display as a light that will automatically turn on when night falls.

This simple project builds on Experiment 3 to monitor the light level and then turn on all the LEDs on the micro:bit's display if the light gets too dim.

What You'll Need

The only thing you need for this project is a micro:bit.

Because the micro:bit will have to stay on for a long time, it's best to use a USB power adapter or Monk Makes Power for micro:bit (see "Powering Your Micro:bit" on page 248) for this project. Batteries run out of power too soon.

Construction

1. Visit the book's code repository at *https://github.com/simonmonk/mbms/* and click the link for **Automatic Night-Light**. Then click **Download** and copy the hex file onto your micro:bit. If you get stuck on this, head back to "Programming the micro:bit" on page 11, where we discuss the process of getting programs onto your micro:bit in full.

2. Once you've programmed the micro:bit, the display should illuminate when you shade the micro:bit with your hand. When you take your hand away, the display should turn off.

Code

The Blocks code for this project is shown here.

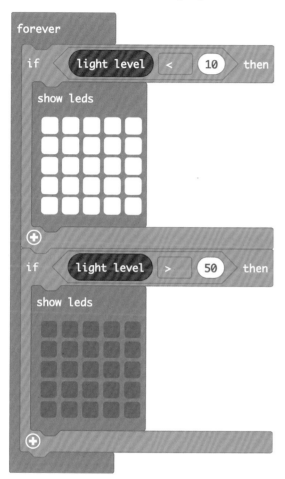

    This code uses two `if` blocks. The first checks whether the light level is less than 10, which would indicate that the environment is very dark. If this condition is met, the code turns on all the LEDs using the `show leds` block with all the LEDs selected.

    The second `if` block turns the LEDs off if the light level is greater than 50. We leave a gap between the two levels of 10 and 50 to make sure the LEDs don't flash on and off when the light is close to either level. The technical name for the difference between on/off values is *hysterysis*.

PROJECT: LIGHT GUITAR

Difficulty: Medium

Now it's time to make a light-controlled guitar! This guitar will play different tones when you wave your hand in front of the micro:bit. You can see a video of this project in action at *https://www.youtube.com/watch?v=OFUYxIYCXQs*.

I recommend attaching your micro:bit, a speaker, and a battery to a guitar-shaped cardboard cutout. (The Mad Scientist is not terribly good at art, so the resemblance in Figure 3-1 is only passing.)

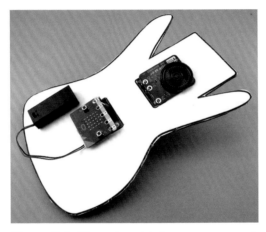

Figure 3-1: The micro:bit becomes a musical instrument.

The cables that connect the micro:bit to the speaker are hidden behind the cardboard to keep things neat.

What You'll Need

For this project, you'll need the following items:

Micro:bit

3 × Alligator clip cables   To connect the micro:bit to the speaker

3V battery pack   To power the micro:bit and speaker (The kind with a switch works best.)

Speaker for micro:bit   To play the sound (Use a Monk Makes Speaker or see Chapter 2 for other speaker options.)

Blu-Tack or self-adhesive pads   To attach the micro:bit to the cardboard cutout

Cardboard   To form the guitar body (You can use a cut-up cardboard box.)

Paper glue (PVA or spray adhesive)   To stick the paper outline to the cardboard

Paper with a guitar outline (optional)   You can either draw your own guitar on paper or cut out and decorate the cardboard directly.

Scissors   To cut out the outline of the guitar

Construction

It's a good idea to create and test the program before attaching your micro:bit to the cardboard, so we'll do that first.

1. Connect the speaker to the micro:bit using the three alligator cables, as shown in Figure 3-2.

    It's a good idea to color-code the cables so you remember which is which. Use black for GND, red for 3V, and any other color for the audio connection from pin 0 of the micro:bit.

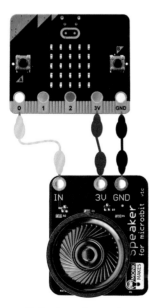

Figure 3-2: Attaching the micro:bit to a speaker

2. Open *https://github.com/simonmonk/mbms/* and click for the **Light Guitar** project. Once the experiment has opened, click **Download**, then copy the hex file onto your micro:bit.

3. Once you've programmed the micro:bit, it should play a different note when your hand gets close to it and shades some of the light. If you don't get a very large range of notes, you may need to adjust its sensitivity (see "Code" on page 71 for how to do this).

4. Disconnect the micro:bit from the USB cable. At *https:// github.com/simonmonk/mbms/*, you'll find a folder called **other downloads** that contains a drawing of a guitar shape in PDF, PNG, and SVG formats. You can draw your own guitar or print out this template.

5. Glue your guitar drawing to your piece of cardboard, as shown in Figure 3-3.

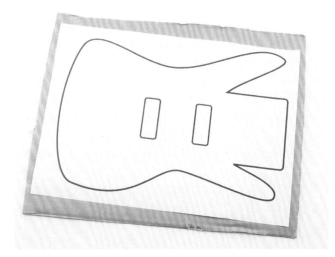

Figure 3-3: Gluing the paper template to the cardboard

6. Using a pair of scissors, cut around the outline of the guitar. Also cut out the two rectangles in the body of the guitar, where you'll put the wires connecting the micro:bit to the speaker. The result should look like Figure 3-4.

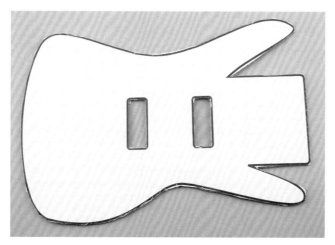

Figure 3-4: Cutting around the template

7.   Using small balls of Blu-Tack, attach the speaker and the micro:bit to the guitar so that the edge connectors of both are accessible from the rectangular slots. Also attach the battery box, as shown in Figure 3-5.

Figure 3-5: Attaching the micro:bit, the speaker, and the battery box

When all three parts are attached to the guitar, the project should look like Figure 3-6.

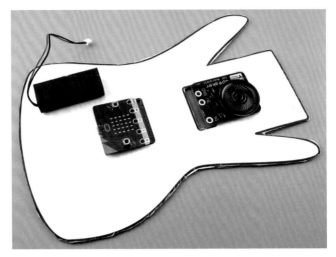

Figure 3-6: The micro:bit, speaker, and battery box have been attached to the guitar.

8. Flip the whole thing over and connect the micro:bit and speaker, as shown in Figure 3-7. Reference Figure 3-2 if you need help with the connections.

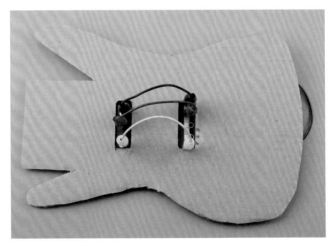

Figure 3-7: Connecting the micro:bit to the speaker

9. Connect the battery box to the battery connector on the micro:bit. You're ready to rock out!

Code

The Blocks code for this project uses arrays.

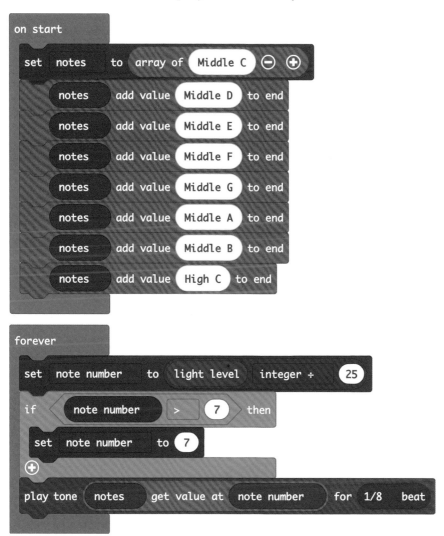

We put the code that needs to be run only when the micro:bit starts up in the on start block. This creates an array of notes. An *array* is a list of values rather than a single value. We're using an array to provide a list of eight possible notes for the micro:bit to play. Different light levels will trigger different notes. The first line inside the on start block defines a new array called notes, which initially just contains the note Middle C. The add value blocks that follow add another seven notes to the array so that the variable notes will eventually contain all eight of the notes we need.

The code inside the forever block will run continuously. This is the code that tells the micro:bit to read the light level and then uses that information to pick a note to play. To do this, the code first reads the light level using the light level block and divides the reading by 25. If you don't hear all eight notes, you may need to adjust the sensitivity by tweaking this value of 25 down a bit.

The set to block names the resulting value note number. Since the note number is the light level divided by 25, it will be a value between 1 and 10.

But setting the maximum light level to 10 would cause problems. When accessing items in an array, you specify the position of the item you want. The maximum position we can specify in this array is 7 (the eight notes are numbered 0 to 7). We want the guitar to work indoors where the light is fairly low. However, if the light is really bright, then note number might turn out to be 10, which would be outside the array. To make sure that we don't exceed this maximum permitted value for note number, we add the if command to check whether the note number is greater than 7. If it is, the code sets it to 7—problem solved.

The play tone block accesses the item in the notes array at the position of the value of note number and plays it for 1/8 of a beat.

PROJECT: INFINITY MIRROR

This seemingly magical mirror (Figure 3-8) is guaranteed to impress any visiting mad scientists. Built into a small picture frame, the mirror looks much deeper than it really is.

Figure 3-8: A micro:bit-controlled infinity mirror

Your micro:bit will control a strip of 30 LEDs mounted inside a picture frame, and we'll put some reflective film on either side of the LEDs, allowing you to create interesting light effects.

You can see a video of this project in operation at *https://www.youtube.com/watch?v=-4Ud47OkIyY*.

NOTE *This project is a little tricky because the reflective film can be difficult to smooth out. If you stick with it, though, the effect is worth it!*

What You'll Need

For this project, you'll need the following items:

Micro:bit

0.5 m Addressable LED Strip (NeoPixels) One WS2812B RGB 5050 SMD Strip should have 60 LEDs per meter and be 0.5 meters long, with self-adhesive backing.

3 × Alligator clip to male jumper cables To connect the micro:bit to the LED strip

Monk Makes Power for micro:bit To power the micro:bit and LED strip. Note that USB power will not provide enough current for this project and a AAA battery pack will provide enough power only if the batteries are very fresh.

6V DC wall wart power adapter To provide enough power for the LEDs. The power adapter should have a DC barrel jack plug on the end. (See "Powering Your Micro:bit" on page 248.)

Deep 7 × 5-inch picture frame See additional information about the picture frame below. You'll need one with plastic spacers, which a deep frame should contain.

Two 7 × 5-inch pieces of reflective window film Samples of reflective window film

Craft knife To cut the reflective film and make a slot in the picture frame spacer for the LED strip's wires

Soapy water To clean the glass and attach the film

Ruler or plastic card To use as a squeegee

Adhesive tape To remove the backing film from the reflective window film

Look for a 7 × 5-inch (17.5 × 12.5 cm) picture frame that has an insert designed for 6 × 4-inch photos. The important thing about the frame is that it has sufficient depth, so look for a frame like the one shown in Figure 3-9a–b. The one I used was 1.25 inches deep. Also make sure that there's glass at the front of the frame. Behind the glass, there should be a plastic spacer (on which we're going to stick the LED strip) and then the backboard.

a.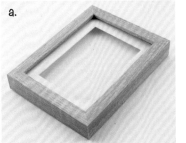 b.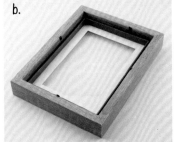

Figure 3-9: A deep photo frame for the infinity mirror: (a) front and (b) with the back removed

The reflective film reflects light, so it acts like a mirror. It normally comes in large rolls for fitting to the windows of a building. Luckily for us, the people who sell this film also usually offer small samples of the material, which are plenty big enough for our needs. You'll find it most easily online. Search on eBay or Amazon by entering the phrase *silver reflective mirror window film* and then look through the listings until you find someone who offers sample sizes.

You're going to need two pieces that are at least 7 × 5 inches, but I recommend getting extra; attaching the film to the glass of the photo frame and removing the bubbles is a little tricky, so it's a good idea to have a few spare sheets in case your first attempts fail.

The easiest place to find the addressable LED strips is on eBay, but it's also sold through hardware sites like Adafruit and SparkFun. Use the search term *WS2812B RGB 5050 LED Strip*. You'll want to buy a strip with 60 LEDs per meter. You'll need half a meter for this project, so 30 LEDs in total. These LED strips come in two varieties: waterproof (for outside use) or with a self-adhesive backing. You need the type with the adhesive backing.

Construction

Now that you have your materials, we'll attach the LEDs and reflective film to the frame. After we program the LEDs, the light will appear to go back into the frame forever, creating the illusion that the frame is deeper than it is.

1. It's worth making sure the LED strip works with your micro:bit before you go through the trouble of constructing the hardware for the project. To test it, connect the LED strip to the micro:bit using the three alligator to jumper cables.

    Note that it's also possible to connect the strip using male-to-male jumper wires and normal alligator clips.

    Also attach the DC power adapter and Monk Makes Power for micro:bit board, as shown in Figure 3-10.

Figure 3-10: Testing the LED strip with a micro:bit

The pins of the alligator to jumper pin cables will fit into the three-way connector attached to the LED strip. Your LED strip may have differently colored wires than the ones shown in Figure 3-10, so it's best to look at the strip itself to see where the wires are soldered to it. One wire will be marked GND; connect this to the GND connection of the micro:bit. A second wire will be marked 5V; connect this to 3V on the micro:bit. The final connection, in the middle, will either be marked DIN (Data In) or have an arrow pointing toward the LED. Connect this to pin 0 of the micro:bit.

2. Got to *https://github.com/simonmonk/mbms/* and click the link for **Magic Mirror**. Click **Download** to download the code and then copy the hex file onto your micro:bit. If you get stuck, head back to "Programming the Micro:bit" on page 11, where we discuss the process of getting programs onto your micro:bit in full.

   If you prefer to use Python, then download the code from the same website. For instructions for downloading and using the book's examples, see "Downloading the Code" on page 34. The Python file for this project is *ch_03_Magic_Mirror.py*.

3. Once you've loaded the program onto your micro:bit, you should see LEDs blink on and off at random.

   Figure 3-11 shows how the infinity mirror is constructed from the picture frame.

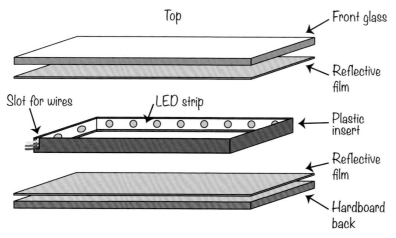

Figure 3-11: How the infinity mirror is constructed

The following steps will lead you through this part of the project build.

4. Take the photo frame apart and remove the plastic insert. Using a craft knife, carefully cut out a slot in one corner of the insert, as shown in Figure 3-12, for the cables to the LED strip to go through.

WARNING   *Craft knives can be very sharp, so be careful and ask an adult for help.*

Figure 3-12: Cutting a groove for the wires in the plastic insert

5. Without removing the backing tape, lay the LED strip around the edge of the plastic insert. If it's not long enough, you can stretch it across the corners, as shown in Figure 3-13. When

you've made sure it fits, peel off the backing film and fix the LED strip to the inside of the insert.

Figure 3-13: Attaching the LED strip to the plastic insert

6. Now set the insert aside. The next step is to attach the reflective film to the glass. Carefully remove the frame's glass and wash it in soapy water. Dry one side and set it down with the wet side up.

    Pick up the reflective film. There will be a backing sheet of transparent plastic on the adhesive side of the film. To separate this thin transparent layer, put adhesive tape on both sides of the film and pull apart (Figure 3-14).

Figure 3-14: Removing the plastic backing from the reflective window film

Place the reflective film adhesive side down onto the wet glass. Using a plastic card or ruler, push the bubbles between the film and glass to the edge of the glass to remove them (Figure 3-15).

Figure 3-15: Removing bubbles from the reflective window film

7. Let the film dry for about an hour. Then flip the glass over and use a craft knife to trim the film to the same size as the glass (Figure 3-16).

Figure 3-16: Trimming the reflective window film

8. Lay the backboard of the frame on your second piece of reflective film and cut the film to the same size as the backing frame (Figure 3-17). You don't need to remove the protective film or stick the film to the board unless the film does not lie flat when you put the frame back together.

Figure 3-17: Trimming the reflective window film for the backboard

9. Now it's time to assemble the frame.

Put the glass back into the frame with the glass side facing out. This will protect the film and make the mirror look better (Figure 3-18a). If your picture frame included a thin piece of cardboard meant to be used with a smaller print, you can insert this after the glass.

Next, put the plastic insert back into the frame, allowing the cable to escape through one corner (Figure 3-18b). If the insert has a wider flat side designed to support the photo, this should face away from the glass.

Finally, place the cardboard back onto the frame, allowing the cable to escape through one corner (Figure 3-18c). Fix the back in place and flip the finished frame over (Figure 3-18d).

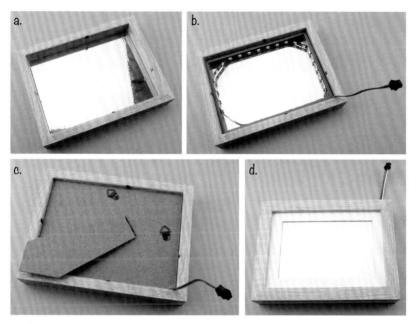

Figure 3-18: Reassembling the frame

10. Reattach your micro:bit, and your magic mirror should spring into life.

Code

We want to program the LEDs to blink on and off at random. To do this, the code picks one of the LEDs at random, then generates a random color and sets the LED to that color.

Blocks Code

Here is the Blocks code for the project.

When using addressable LEDs (called NeoPixels in the Blocks code), you need to assign the `NeoPixel at` block to a variable, which we have called `leds`, in your `on start` block. You must supply the `NeoPixel at` block with the pin used to control the LEDs. In this case, we'll use pin P0. Then specify the number of LEDs on the strip, which is 30 in our case. Finally, tell `NeoPixel at` how to define the colors. In this case, that's the standard RGB format.

We again use a `forever` block to keep this code running continuously. In the `forever` block, we start by generating a random number between 0 and 29 using the `pick random` block and assign this number to the variable `led`. This selects one of the LEDs. The `color` variable is then assigned random amounts (between 0 and 255) of red, green, and blue.

The `show` block updates the LED strip with the change we've just made.

Try experimenting with the code to change the LED colors.

MicroPython Code

Here's the MicroPython version of the code.

```
from microbit import *
import neopixel, random

leds = neopixel.NeoPixel(pin0, 30)
```

```
while True:
    led = random.randint(0, 29)
    color = (random.randint(0, 255), random.randint(0, 255),
random.randint(0, 255))
    leds[led] = color
    leds.show()
    sleep(5)
```

This code works in much the same way as the Blocks version. Notice that you have to include the neopixel and random libraries at the top of the program so that you can access the LED strip and generate the random numbers.

The MicroPython neopixel library first defines the LED strip as using pin 0, which is the micro:bit pin we attached it to, and having a length of 30 LEDs. It does this using the command leds = neopixel.NeoPixel(pin0, 30).

The while loop then does the same job as the forever block in the Blocks version, picking an LED at random and setting it to a random color.

How It Works

Figure 3-19 shows how the infinity mirror works.

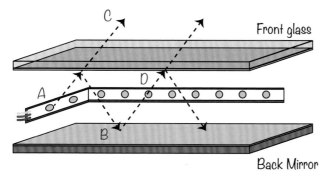

Figure 3-19: Light bouncing around in the infinity mirror project

Light from the LEDs heads off in all directions (A in Figure 3-19). Some of that light travels up, toward the front of the frame. The mirror film reflects some—but not all—of this light back down to the mirror at the back of the frame (B). The rest of the light escapes from the frame (C), and some of it finds its way to the back of your eye, where it forms an image.

Meanwhile, some of the light that found its way down to the mirror (B) will bounce back up toward the front glass again (D). Some of this light will escape the picture frame entirely and find its way to your eye, and the rest will perform yet another bounce. It's this bouncing back and forth of the light that causes an endless succession of LEDs to seem to disappear into the mirror, getting dimmer each time a proportion of the light is lost.

SUMMARY

In this chapter, you have learned how to both measure and make light with your micro:bit. In Chapter 4, the Mad Scientist turns their attention to magnetism.

4

MAGICAL MAGNETISM

As you saw in Chapter 1, the micro:bit has a built-in magnetometer that serves variety of purposes. In this chapter, we'll use it to turn your micro:bit into a compass that tells you which direction you're facing. We'll also measure the magnetic fields of neodymium magnets. Then, we'll make a magnetic alarm that rings whenever someone opens the door.

PROJECT: COMPASS

Difficulty: Easy

In this project, shown in Figure 4-1, you'll use the micro:bit's built-in magnetometer as a compass. It will display an arrow on the screen that points toward magnetic north.

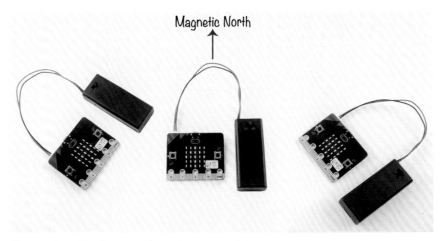

Figure 4-1: Finding north with a micro:bit compass

Unlike a conventional compass, this compass doesn't always point north. Instead, it shows you which way to turn in order to be facing north.

Rotate until the arrow is facing straight ahead, and you're looking north!

What You'll Need

For this experiment, you will need:

Micro:bit

Battery pack

You only need a micro:bit to make this compass, but if you plan to navigate outdoors, then you'll need a battery pack as well.

Construction

1. Go to *https://github.com/simonmonk/mbms/* to access the book's code repository and click the link for **Compass**. Once the program has opened, click **Download** and then copy the hex file onto your micro:bit. If you get stuck on this section, head back to Chapter 1, where the process of getting programs onto your micro:bit is explained in full.

    If you prefer to use Python, then download the code from the same website, along with instructions for downloading and using the book's examples. The Python file for this experiment is *ch_04_Compass.py*.

2. Once you've loaded the program onto your micro:bit, you'll see a message prompting you to move your micro:bit in a certain way. Follow the instructions. This will happen every time you flash a program onto the micro:bit that uses the magnetometer.

    The purpose of moving your micro:bit this way is to calibrate its magnetometer. The magnetometer chip on the micro:bit is sensitive to local magnetic field variations. By moving the magnetometer in different directions, you help its internal filtering software to compensate for the local distortions to Earth's magnetic field. This is why it's a good idea to calibrate your compass again if you take it outside, away from the metal objects commonly found indoors. Also, a battery pack can affect the magnetometer, so it's best to calibrate with the same hardware you'll use on the final project.

Calibrating your micro:bit's magnetometer is a bit like doing a puzzle. As you move your micro:bit, more LEDs will light up (Figure 4-2).

Figure 4-2: Calibrating your micro:bit's magnetometer

3. To use the compass, attach a battery pack to your micro:bit and take it outside, well away from items such as computers and household appliances. Hold the micro:bit level (so it's flat). If the arrow points to the right or left, slowly turn in that direction until the arrow points straight ahead. When it does, you're facing magnetic north.

    If the compass isn't pointing north, try calibrating it again by pressing **button A**.

Code

The Blocks and MicroPython versions of the code follow the same pattern. The *heading*, or *bearing*, (the direction the micro:bit is pointing) is measured and then the program decides which of the three arrows (north, west, or east) it should display in order to direct the user in the right direction.

Blocks Code

Here is the Blocks code for this project.

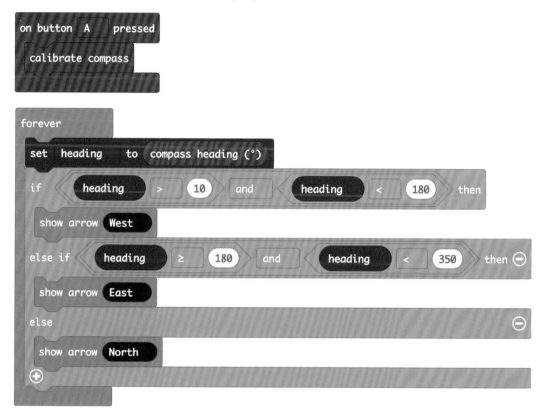

In this project, we'll set up the code so that you can calibrate the magnetometer any time you want by pressing button A.

The `calibrate compass` block is inside the `on button A pressed` block. The `calibrate compass` block will display the dot we used to calibrate the magnetometer earlier. You can then roll it around the display, as we did at the start of this project.

In the forever block, assign the value of compass heading (degrees from 0 to 359) to a variable called heading, which will contain this heading.

Then create a large if block that consists of an if, an else if, and an else. This big block will test which direction the micro:bit is facing and display an arrow on the LED display showing you which way to turn to be facing north.

If you're writing this code yourself, you will notice that the Blocks code library seems to offer only two kinds of if blocks: if then and else. To add the else if then section, click the + circled in Figure 4-3.

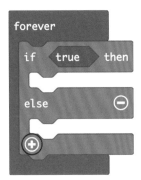

Figure 4-3: Adding another section to if blocks

The first line of the large if block tests whether heading is between 10 and 180 degrees, where 0 degrees is north. Thus, a heading value between 10 and 180 would indicate that you were facing east. In this case, the micro:bit would display the west arrow, which points left. This arrow tells you to turn to the left in order to face north.

If heading is not between 10 and 180, then the else if part of the block tests whether heading is greater than or equal to 180 and less than 350 degrees. If it is, then the micro:bit displays the east arrow, which tells you to go right.

If heading is between 350 and 10 degrees, the else part of the if block will display the north arrow, which points forward—you're on the right track! Figure 4-4 shows how this works (the numbers are possible values of heading).

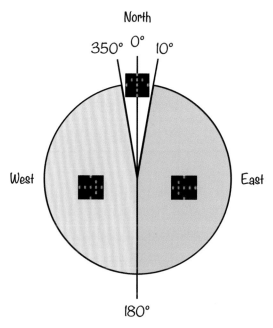

Figure 4-4: The arrow will point in one of three directions, guiding you north.

MicroPython Code

Here is the MicroPython code for the project:

```
from microbit import *

while True:
    heading = compass.heading()
    if heading > 10 and heading < 180:
        display.show(Image.ARROW_W)
    elif heading >= 180 and heading < 350:
        display.show(Image.ARROW_E)
    else:
        display.show(Image.ARROW_N)

    if button_a.was_pressed():
        compass.calibrate()
```

This code uses the same logic as its Blocks code equivalent: it checks the reading of the magnetometer and tells the micro:bit to display an arrow. Again, if the heading is between 10 and 180 degrees from north, the micro:bit shows a west arrow; if it's between 180 and 350 degrees, the micro:bit displays an east arrow; and if the heading is within 20 degrees of north, an arrow points straight up, telling you to keep going in the same direction.

Things to Try

See whether you can use a magnet to confuse the compass about which direction is north.

How It Works: The Earth's Magnetic Field

Earth's magnetic north and south poles have strong charges, creating a magnetic field all around the globe, as shown in Figure 4-5.

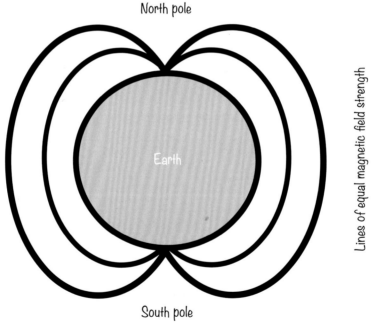

Figure 4-5: Earth's magnetic field

The micro:bit's magnetometer measures the strength of Earth's magnetic field to deduce the direction the micro:bit is facing.

Interestingly, the magnetic north pole is not exactly at the top of the spinning ball that is Earth. The magnetic field can be centered up to 20 degrees away from the geographic poles, depending on where you are in the world.

Another interesting fact about the magnetic poles is that they're shifting at a rate of about 6 miles (10 km) a year. One day, the poles will flip completely, meaning that magnetic north will be at the southern end of the globe. This process happens on Earth about every 450,000 years.

EXPERIMENT 4: MEASURING MAGNETIC FIELDS

Difficulty: Medium

As we just saw, the micro:bit's built-in magnetometer is sensitive enough that you can use it as a compass to find north. It's also a great tool for measuring the strength of a nearby magnet. In this experiment (shown in Figure 4-6), you'll measure the strength of a magnetic field at various distances.

What You'll Need

In this experiment, you'll need the following:

Micro:bit

Strong neodymium magnet

Ruler (preferably showing millimeters)

You'll find all sorts of shapes and sizes of magnets on eBay. As you can see in Figure 4-6, I'm using a disk-shaped one with a diameter of 10 mm, but any neodymium magnet of a similar size should work. Note that I've put a bit of tape labeled *N* on one side of the magnet (more on this later).

Figure 4-6: The neodymium magnet used for Experiment 4

WARNING   *Neodynmium magnets are very strong, so be careful when handling them. If two of them get stuck together, it can be hard to separate them.*

Construction

The program repeatedly scrolls a number across the micro:bit's display. This number represents the overall magnetic field strength detected by the magnetometer. We are going to measure the field strength at various distances from the magnet.

1. Go to *https://github.com/simonmonk/mbms/* to access the book's code repository and click the link for **Experiment 4: Magnetic Fields**. Once the program has opened, click **Download** and then copy the hex file onto your micro:bit. If you get stuck on this, head back to Chapter 1, where the process of getting programs onto your micro:bit is explained in full.

    If you prefer to use Python, download the code for this project from the same website, along with instructions for downloading and using the book examples. The Python file for this experiment is *Experiment_04.py*.

2. Set your magnet on a flat surface so that it's balanced on its curved edge. A neodymium magnet is so strong that if you set it down in this way, it will align itself with Earth's magnetic field (by spinning until one side is facing north),

just like a compass needle. Allow your magnet to do this and then put a piece of tape on the side that's facing north. (You might need to use the Compass project, described earlier in this chapter, to find north so you don't accidentally mark south instead.)

3. Align your ruler so the 0 cm mark is pointing north and the 30 cm mark is pointing south, as shown in Figure 4-7. The ruler allows you to make sure the magnet continues to face north so that the effect of Earth's magnetic field remains constant.

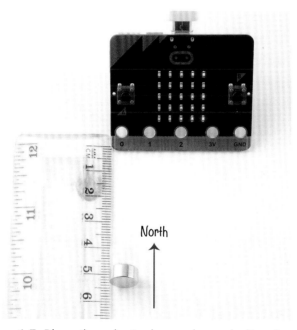

Figure 4-7: Place the ruler in the north-south direction to make sure your magnet continues to face north.

4. Set the magnet on its curved edge next to the 1 cm mark on the ruler. Make sure the north side is facing the micro:bit. Point the micro:bit south and align its edge with the 0 cm mark. Record the reading displayed on the micro:bit in the following table, in the 1 cm row. The unit of magnetic field strength is a *tesla* (which is a very, very strong magnetic field), so the readings from the micro:bit are in *microtesla* or millionths of a Tesla, represented by µT).

Distance from micro:bit to magnet (cm)	Magnetic field strength (µT)
1	
2	
3	
4	
5	
6	
7	
8	
9	
10	

5. Move the magnet so it's next to the 2 cm mark on the ruler. Enter the new number displayed on the micro:bit in the corresponding row in the table. Repeat this process for all the readings up to 10 cm. Notice that the field strength decreases quickly as you move the magnet away from the micro:bit.

6. Once you've taken all the readings, try plotting a chart of your results. Make the vertical axis, or *Y axis*, show the field strength in µT and plot the distance of the magnet from the micro:bit in cm against the horizontal axis, or *X axis*.

    You can plot this graph by hand on paper, or, if you prefer, you can make a copy of the Google Sheets spreadsheet I've created for this experiment and then replace my readings with the numbers you have just written down. You will find a link to this spreadsheet at *https://github .com/simonmonk/mbms/*. Open this link and, from the spreadsheet menu, select **File ▸ Make a Copy** to make your own copy. Figure 4-8 shows the completed spread-sheet with a chart of the readings underneath it.

Note that your readings might be different, as your magnet may be weaker or stronger.

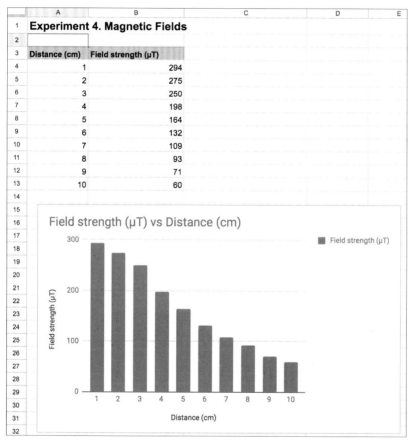

Figure 4-8: Make a graph of your results.

We'll discuss these results in "How It Works: The Strength of Magnets" on page 99.

Code

The code for this experiment is simple. It takes a reading, displays it, and then repeats.

Blocks Code

Here is the Blocks code for this project.

The show number block displays the result of the magnetic force block inside the forever block.

The magnet force block has four options, which you can see in the drop-down menu of this block. You can either get a separate reading for the force in one of the three dimensions X, Y, or Z, or you can get the combined force measurement from all three dimensions. We will use the combined measurement, which is the strength option in the drop-down menu.

MicroPython Code

Here is the MicroPython version of the code:

```
from microbit import *
while True:
    display.scroll(str(int(compass.get_field_strength() / 1300)))
```

In MicroPython, the get_field_strength function takes readings from the micro:bit's magnetometer. Unlike the magnetic force block of the Blocks code, the MicroPython code does not specify the units for the measurement it returns, but the reading we get back is about 1,300 times the reading reported by the equivalent magnetic force block. Therefore, divide the number by 1,300 and then convert this value with the int function to make it an integer.

Use the str function to convert the number to a text string so the scroll function can display it on the micro:bit screen.

Things to Try

Using the program for Experiment 4, try moving the magnet farther and farther away to see how far away the magnet can be and still show up on your micro:bit.

How It Works: The Strength of Magnets

The chart in Figure 4-8 shows that as you move the magnet away from the micro:bit, the magnetic field strength falls very rapidly at first and then gradually levels off. In fact, the relationship between magnetic field strength and distance from the magnet follows something called the *inverse square rule*. That is, the field strength is proportional to 1 divided by the square root of the distance from the magnet.

What this means is that when the distance between the magnet and the micro:bit doubles, the field strength falls to a quarter of its previous strength.

The strength of Earth's magnetic field when measured on the ground ranges from 25 to 65 µT. When you measure the magnetic force without a magnet nearby, you should get readings in this range. Even the super-strong neodymium magnet used in this experiment has a field strength of only a few hundred µT at a distance of a few centimeters. Magnets that are used to look inside human bodies during magnetic resonance imaging (MRI) scans typically have a field strength (where the person lies) of 0.5 to 1.5 T. That's several thousand times stronger than the neodymium magnet. This is why you have to take off any metal you are wearing when you go into an MRI scanner!

PROJECT: MAGNETIC DOOR ALARM

Difficulty: Medium

We're going to create a door alarm! By attaching a magnet to a door and a micro:bit to a door frame, the Mad Scientist can be alerted to the arrival of guests. The micro:bit will sound an alarm when it detects a change in the strength of the magnetic field. This is exactly what happens when someone opens the door, moving the magnet farther away.

Figure 4-9 shows the project attached to the Mad Scientist's door.

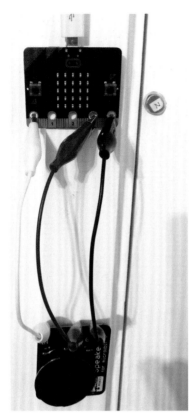

Figure 4-9: A magnetic door alarm

What You'll Need

For this project, you will need the following items:

Micro:bit

3 × Alligator clip cables   To connect the micro:bit to the speaker. These cables should be at least 15 cm (6 inches) long.

USB power adapter or switched 3V battery pack To power the micro:bit and speaker. See the appendix for other options for powering your micro:bit.

Speaker   You can use any of the methods for getting sound out of a micro:bit discussed in Chapter 2. Here, I used a Monk Makes Speaker for micro:bit.

Adhesive putty or self-adhesive pads To attach the micro:bit and speaker to the door frame and the magnet to the door

Neodymium magnet As used in Experiment 4

Construction

Make sure to test the project before you install it on your door, since you'll need to experiment a little to determine the right field strength to use. You'll also need to calibrate the magnetometer before sticking it to a door frame.

1. Connect the speaker to the micro:bit using the three alligator cables, as shown in Figure 4-9.

   It's a good idea to color-code your cables by using black for GND, red for 3V, and any other color for the audio connection from pin 0 of the micro:bit. That way, you won't forget which cable is which. You'll need to place the speaker at least 15 cm away from the micro:bit, as speakers have strong magnets that can disrupt the micro:bit's magnetometer.

2. Go to *https://github.com/simonmonk/mbms/* to access the book's code repository and click the link for **Magnetic Alarm**. Once the program has opened, click **Download** and then copy the hex file onto your micro:bit. If you get stuck on this, head back to Chapter 1.

   If you prefer to use Python, you'll find the code for this at the same website, along with instructions for downloading and using the book's examples. The Python file for this experiment is *ch_04_Magnetic_Alarm.py*.

3. Once you have programmed the micro:bit, when you put the magnet close to the micro:bit, the speaker should be silent. Then when you move the magnet more than a few inches away, the alarm should sound. Once the program is working, disconnect the micro:bit from your computer and attach it to either a USB power adapter or battery pack. Then attach the micro:bit and speaker to your door frame. Fix the magnet to the door itself, within an inch (2.5 cm) of the micro:bit.

WARNING *Adhesive putty can make a mess, so make sure you get permission before you start sticking things to the door and frame.*

If you're using a USB power adapter, you'll need an electrical outlet close enough to the door for the USB power adapter to reach.

Code

In both versions of the code, the magnetic field strength is first read and then compared with a threshold. If the threshold is exceeded, then a note is played.

Blocks Code

Here is the Blocks code for the project.

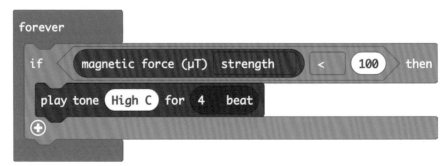

This code is basically saying that if the micro:bit reads a magnetic force strength of less than 100 µT, it should play a particular note for 4 beats. Since the strength of Earth's magnetic field at the surface has a range of 25 to 65 µT, setting a threshold at 100 µT should ensure we're detecting the magnet and not just picking up Earth's magnetic field.

MicroPython Code

Here's the MicroPython version of the program:

```
from microbit import *
import music

while True:
    if compass.get_field_strength() < 160000:
        music.pitch(523, 1000)
```

This code works in the same way as the Blocks code, but because the units of get_field_strength are different from those of the magnetic force block, the threshold for sounding the alarm is different than in the Blocks code version. The units of the field strength returned by get_field_strength are not specified in its documentation (*https://microbit-micropython.readthedocs .io/en/latest/compass.html*), so I chose the threshold value of 160,000 by trial and error.

We supply the command music.pitch with a frequency of tone to produce. Unlike with the Blocks code, here we have to specify the frequency of the sound wave rather than a note. A frequency of 523 hertz (Hz) is the frequency of high C. The music.pitch command also requires a length of time to play the note. Here we've told the micro:bit to play the note for 1,000 milliseconds, or 1 second.

Things to Try

Could a thief defeat your new alarm? Try placing a second magnet on the other side of the door frame, opposite the micro:bit. If the second magnet is close enough to the micro:bit, you may be able to open the door without triggering the alarm.

You could also change the alarm tone by changing the frequency to some other value. The following table shows some notes and their corresponding frequencies (to the nearest whole number).

Note	Frequency (Hz)
Middle C	262
D	294
E	330
F	349
G	392
A	440
B	494
High C	523

You can find a full table of musical notes and their frequencies at *https://www.liutaiomottola.com/formulae/freqtab.htm*.

SUMMARY

The micro:bit's magnetometer is useful for more than working out which direction you're facing. In this chapter, we've explored how you can use it to sense the presence of a magnet and even to do some science by measuring magnetic field strength.

In the next chapter, we'll explore one of the micro:bit's other built-in sensors—its accelerometer.

5

AMAZING ACCELERATION

The accelerometer is arguably the most useful of the BBC micro:bit's built-in sensors. It lets you measure the direction and strength of a force, such as gravity, that is acting on the micro:bit.

You can use the accelerometer for many things, including:

- Detecting gestures like shaking, or even finding out that your micro:bit is falling
- Learning which way and how much the micro:bit is tilted and using this to control your micro:bit
- Measuring how quickly your micro:bit is accelerating when it moves. (For example, you might use it in a pedometer to see how many steps you take in a day.)

 # EXPERIMENT 5: GESTURES

The micro:bit's software includes gesture recognition, so it can respond to certain motions, like tilting or shaking, that are picked up by the accelerometer. In this section, you'll program the micro:bit to display a smile whenever you shake it.

Later in this chapter, you'll learn how to deal with the raw data that comes straight from the accelerometer chip so you can measure acceleration.

What You'll Need

In this experiment, you'll just need your micro:bit.

Construction

1. Go to *https://github.com/simonmonk/mbms/* to access the book's code repository and click the link for **Experiment 5: Gestures**. Once the program has opened, click **Download** and then copy the hex file onto your micro:bit. If you prefer to use Python, then download the code from the same website. For instructions for downloading and using the book's examples, see "Downloading the Code" on page 34. The Python file for this experiment is *Experiment_05.py*.

2. Once the program starts, shake your micro:bit, and you should see a smile appear, then disappear, on the display.

Code

Using gestures in your program is similar in both Blocks and MicroPython. Both languages have the same set of gesture types. The main difference is that in MicroPython there is no event mechanism for handling events; instead, you have to keep checking for gestures in a loop.

Blocks Code

Here is the Blocks code for this experiment.

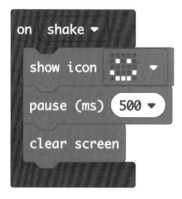

The acceleration gestures available in the Blocks code work as events, just as pressing a button does. Start with an on block. Click the triangle in to access a drop-down menu of possible gestures, shown in Figure 5-1, and select shake.

Figure 5-1: Selecting different gestures

THE DIFFERENT GESTURES

The shake gesture responds to any significant shaking of your micro:bit. The next two gestures, logo up and logo down, respond to tilting the micro:bit front to back. The logo this code refers to is the icon printed on the micro:bit board, near the USB socket.

The screen up and screen down gestures respond to the micro:bit's orientation. For example, if you placed it screen side down, you'd trigger the screen down event.

The tilt left and tilt right events respond when you tilt the micro:bit from side to side by more than about 60 degrees. You have to tilt it quite a lot to trigger these events.

The final four events relate to the overall force acting on the accelerometer rather than the force in any particular direction. For example, if the micro:bit is in free fall, then the fall will trigger the free fall event. The gestures 3g, 6g, and 8g detect different amounts of force acting on the accelerometer, measured in *g* (the acceleration due to gravity, or the *g-force*). For example, you could detect a finger tap on the micro:bit. Tapping the micro:bit doesn't move it much, so you might think that not much force is involved, but in fact tapping can exert quite a high g-force.

MicroPython Code

Here's the MicroPython code for this experiment:

```
from microbit import *

while True:
    if accelerometer.was_gesture('shake'):
        display.show(Image.HAPPY)
        sleep(500)
        display.clear()
```

You'll recognize most of the code here. We import the usual library and then start a `while True` loop so the main code runs continually. This loop checks whether the micro:bit has detected shaking and, if it has, shows the happy face!

Because MicroPython doesn't have the concept of "events," you have to use the `was_gesture` function inside a `while` loop to check for the shake gesture. You can also replace the `shake` block with `up`, `down`, `left`, `right`, `face up`, `face down`, `freefall`, `3g`, `6g`, or `8g`.

Things to Try

Try adding more gestures to the program for Experiment 5. You can even make each gesture trigger a different icon on the screen.

In Chapter 10, you'll use gesture detection again to steer a micro:bit-controlled robot rover!

How It Works: Force, Acceleration, and Gravity

We've been talking about force, acceleration, and gravity as if the accelerometer measures all of these things, but it really just measures the distance a certain mass moves. Then it computes the other measurements. Let's take a look at what the accelerometer chip actually does to get a better idea of what these three terms mean.

Figure 5-2 shows a rough sketch of what's inside the tiny accelerometer chip attached to your micro:bit.

Imagine a tiny mass *m* attached to a spring. (The mass is drawn as a ball in Figure 5-2, but its shape isn't important.) Normally, the mass is in position A, but if it's being pulled against the spring by some force (for example, gravity), then it will move to some other position, which we'll call position B. The stronger the force, the bigger the difference between A and B. By measuring this distance, the accelerometer calculates the force acting on the mass.

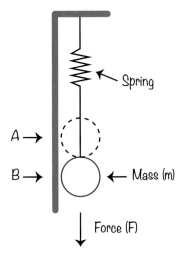

Figure 5-2: The micro:bit's accelerometer

Looking at Figure 5-2, you can see that the force of gravity alone will pull the mass down. But the more the spring is stretched, the more the spring pulls the mass in the opposite direction to gravity. So gravity applies downward force to the mass, stretching the spring, which then applies upward force. When the two forces are equal but acting in opposite directions, the difference between the distances B and A is a measure of the gravitational force acting on the micro:bit. The larger the distance between A and B, the greater the force acting on the mass.

Remember that the g-force relates to the acceleration that occurs due to gravity. Acceleration is an increase in speed. In other words, if you drop your micro:bit off a tall building, the acceleration *g* is the number of meters per second the micro:bit's speed increases by every second. If you dropped your micro:bit off a tall building in a vacuum, there wouldn't be any air to slow down the micro:bit, and in this case, the speed of the micro:bit would increase by about 9.8 meters per second for every second it fell. So, if it were to start at a speed of 0 (as you're holding it in your hand) and then fall for 10 seconds, it would reach a speed of 98 meters per second (about 220 miles per hour).

Also, when the micro:bit meets the ground after traveling at 220 miles per hour, it will probably be smashed to bits.

However, if you dropped your micro:bit in a vacuum, then its accelerometer would read 0 even though it was clearly accelerating at 9.8 meters a second. This is because the accelerometer is not really measuring acceleration. As you saw in Figure 5-3, it's measuring the force acting on the mass inside of the micro:bit. That force would be zero if the mass and the spring were accelerating at the same rate, which they would be since they're both in the micro:bit.

If you were in space, well away from the gravitational pull of celestial bodies, then the acceleration would be equal to the force acting on the object divided by the mass of the object. Because the mass is always the same, the accelerometer can tell us the acceleration of the micro:bit as long as some force, any force, is acting on it.

The accelerometer chip is actually more advanced than Figure 5-2 suggests because it has three force-measuring devices in it, set to measure force in three directions, all at right angles to each other. In other words, the chip measures the force acting on it in three dimensions, X, Y, and Z, as shown in Figure 5-3.

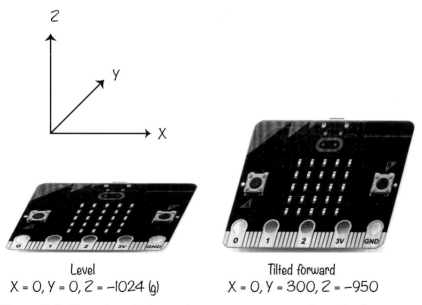

Level
X = 0, Y = 0, Z = –1024 (g)

Tilted forward
X = 0, Y = 300, Z = –950

Figure 5-3: The micro:bit's accelerometer measures force in three dimensions.

If your micro:bit were sitting flat on a table, the X dimension would run left to right, the Y dimension would run front to back, and the Z dimension would run above and below the tabletop. So, if the micro:bit is completely horizontal on a flat tabletop, the force of gravity will only be acting on the Z (up-down) dimension, while the X and Y dimensions will measure 0 acceleration. Now, if you tilt the micro:bit forward a little, then some of the force due to gravity will act on the Y dimension, so the Y value will no longer be 0. This also means that slightly less gravitational force is acting on the Z dimension, so the value of Z will decrease slightly.

From the description so far, you might think that the accelerometer is only useful for measuring the force of gravity. In fact, looking at Figure 5-2, it is easy to imagine how shaking the micro:bit or causing any kind of acceleration on it would change the position of the mass.

 # EXPERIMENT 6: REAL-TIME ACCELERATION PLOTTING

Mu has a great feature that will plot data coming from your micro:bit in real time. In this experiment, you'll use the Plotter feature to see how the acceleration data changes when you move your micro:bit.

NOTE *At the time of writing, the Plotter feature is available only on the Windows and Mac versions of the Mu Editor. The experiment also shows you how to get an overall measurement of acceleration, as well as separate readings for each of the X, Y, and Z dimensions.*

What You'll Need

For this experiment, you just need a micro:bit connected to your computer by a USB cable.

Construction

1. This project uses Mu's Plotter feature, which you'll need Python for, so there's no Blocks code for this. Find the code at *https://github.com/simonmonk/mbms/*. The Python file for this experiment is *Experiment_06.py*. Flash the program onto your micro:bit.

2. Click the **Plotter icon** in the Mu toolbar to open Mu's Plotter, shown in Figure 5-4. If you want to see the raw data that Mu is using to create graphs, click the **REPL** button.

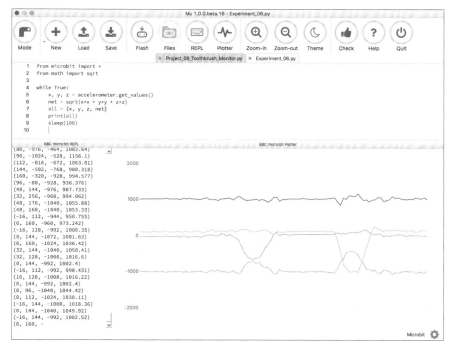

Figure 5-4: Plotting accelerometer data using Mu

3. Tilt the micro:bit this way and that to see the changes in the Plotter.

As you can see in Figure 5-4, there are four plots, each drawn in a different color. There is one plot for each dimension (blue for X, green for Y, and orange for Z). There is also a purple plot for the *net acceleration*, which is the combination of the forces in all three dimensions. We'll explain how the net acceleration is calculated in "How It Works: Calculating Net Acceleration" on page 11.

If you find that readings don't appear in the REPL area of the screen, you may have an old micro:bit that needs an update for this to work. If this is the case, follow the instructions at *https://support.microbit.org/support/solutions/articles/19000019131-how-to-upgrade-the-firmware-on-the-micro-bit/* to update your device.

Code

Here's the Python code that sends the accelerometer data to Mu's plotter:

```python
from microbit import *
from math import sqrt

while True:
    x, y, z = accelerometer.get_values()
    net = sqrt(x*x + y*y + z*z)
    all = (x, y, z, net)
    print(all)
    sleep(100)
```

Import the usual micro:bit library. Then import the square root function `sqrt` from the Python math library. We'll use this function to calculate net acceleration.

Add a `while` loop, which gets the readings for the X, Y, and Z dimensions from the accelerometer in one go using the get_values method. This line will return a *tuple*, which is a data structure capable of holding multiple values. Assign this tuple to three variables: x, y, and z. These will hold the three values, respectively. We use these variables to calculate the net acceleration, which we assign to the variable net.

Next, send the net acceleration value along with the individual values for the X, Y, and Z accelerations to Mu for plotting. Mu expects the values you want to plot to be in the form of a tuple, so create a new tuple, all, that contains all four values.

Finally, print the tuple, which not only shows the values on Mu's REPL (for more information on this, see "The REPL" on page 23) but also provides these values to the Plotter. You can see the values printed in the bottom left of Figure 5-6.

How It Works: Calculating Net Acceleration

To calculate the net acceleration on the X, Y, and Z dimensions, you need to use some ancient Greek technology: the *Pythagorean theorem.*

To understand how this works, picture an acceleration force as a line with an arrow on the end. The arrow indicates the direction of the force, and the length of the line indicates how strong the force is. Lines like these are called *vectors*, and they are used often in physics.

Vectors are easier to understand in two dimensions than in three. Figure 5-5 shows some two-dimensional vectors on the X and Y dimensions. The blue vector along the X axis has a strength of 4, and the green vector along the Y axis has a strength of 3.

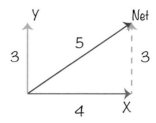

Figure 5-5: Two-dimensional vectors

The purple vector, equivalent to the combined effect of the green and blue vectors, is the *net vector* our program calculated earlier. The net vector is useful because its length tells you the overall strength of the force. To calculate the strength of the net vector, you can use the Pythagorean theorem.

The Pythagorean theorem says that in a right triangle (a triangle that has a right angle), the square of the *hypotenuse* (the triangle's longest side) equals the sum of the squares of the other two sides.

Looking at Figure 5-5, you can see that we do indeed have a right triangle, because the blue and green vectors intersect at a right angle. Using the Pythagorean theorem, we can say that the square of the length of the purple line is equal to $3^2 + 4^2$. That's 9 + 16, which equals 25. So, the length of the purple line is the square root of 25, which is 5.

The math works in three dimensions as well as two. To find the length of a net vector for the dimensions X, Y, and Z, take the square root of the sum of the squares of all three X, Y, and Z vectors.

Figure 5-6 shows the three forces in the X, Y, and Z dimensions at some point in time.

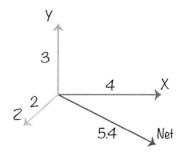

Figure 5-6: Acceleration forces as vectors in three dimensions

Here's the line of code that calculates the length of the single vector that would replace the X, Y, and Z vectors.

```
net = sqrt(x*x + y*y + z*z)
```

The X, Y, and Z values are found in a[0], a[1], and a[2], respectively, in the tuple returned by accelerometer .get_values().

If the X, Y, and Z vectors were of lengths 4, 3, and 2, as shown in Figure 5-6, then calculating the square of the length of the net vector would look like this:

$$4 \times 4 + 3 \times 3 + 2 \times 2 = 16 + 9 + 4 = 29$$

So, the length of the vector would be the square root of 29, or about 5.4.

PROJECT: TOOTHBRUSHING MONITOR

Difficulty: Easy

The Mad Scientist is usually a bit distracted by all the interesting experiments going on, so they need some help brushing their teeth thoroughly. That's why the Mad Scientist has used a micro:bit to create a toothbrush monitor (Figure 5-7) that counts the number of strokes of the toothbrush.

Figure 5-7: A toothbrush monitor

The toothbrush monitor displays a score that ranges from 0 to 9 on the display. It increases the score by 1 for every 50 strokes of the toothbrush. When it counts a score of 10, it displays the happy face icon to show the Mad Scientist that toothbrushing is all done—until next time.

WARNING   *Don't let your micro:bit get wet—it might break!*

What You'll Need

For this project, you will need the following items:

Micro:bit

3V battery pack   To power the micro:bit. (The switched type battery box is preferred.)

Manual toothbrush   Not the electric type

2 × Elastic bands   To attach the micro:bit and battery pack to the toothbrush

This project is intended for a manual toothbrush. Electric toothbrushes won't work with this project.

Construction

1. Go to *https://github.com/simonmonk/mbms/* to access the book's code repository and click the link for **Toothbrushing Monitor**. Once the program has opened, click **Download** and then copy the hex file onto your micro:bit.

    If you prefer to use Python, then download the code from the same website. For instructions for downloading and using the book's examples, see "Downloading the Code" on page 34. The Python file for this experiment is *ch_05_Toothbrush_Monitor.py*.

2. Plug the battery pack into the micro:bit and attach the battery and micro:bit to the toothbrush using elastic bands, as shown in Figure 5-7.

    When positioning the bands, make sure they don't hide the display and that you can still reach the battery pack's on/off switch. Also, check that the bands aren't over the reset switch on the back of the micro:bit.

3. Switch on the battery pack. Once it's on, the micro:bit should show 0. Brush vigorously, and after a little while, the display should show 1.

Code

The programs for this project measure the acceleration and, if it is above a certain level, they add 1 to a count variable to record the number of brush strokes.

When the count of brush strokes becomes big enough to qualify as another point in the tooth-brushing score, then the score is also incremented and displayed.

Eventually, when the score gets to 10, the smiley face icon is shown on the micro:bit's display.

Blocks Code

Here is the Blocks code for the project.

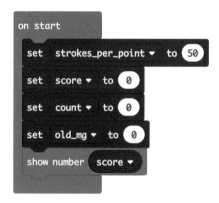

This is probably the most complex program so far in the book. The on start block defines four variables. Here's what each of them does:

strokes per point   This specifies the number of brush strokes needed to advance your score by a point. If you are a lazy brusher, then you can decrease this number so that the points increase more quickly and you get your smiley-face reward faster. Note, however, that your next visit to the dentist may not be fun.

score   This is the number that increases as you complete each set of 50 brush strokes until it reaches 10.

count   This is used to track the number of strokes completed since you earned a point. It starts at 0 and resets each time the score goes up.

old mg   This variable holds the acceleration value in milli-gravities the last time it checked for a brush stroke. The program will compare this value to the new value to detect brush strokes.

We make a forever block, and inside we have a set XX to block. From the drop-down, we select **mg** so that that this block puts the overall acceleration reading into a variable called mg. In the acceleration (mg) block's drop-down menu, choose **strength,** which does the Pythagorean theorem calculation for you. (If you're using MicroPython, you'll have to calculate it yourself.)

Then a set change in mg block calculates the change in acceleration strength by subtracting the current strength from the previous strength (held in old mg). If the change is greater than 800—indicating the start of a brushing movement—the count increases by 1. The value of 800 was chosen by looking at Figure 5-8, which is a plot of the net acceleration for a short period of vigorous toothbrushing.

Figure 5-8: A plot of net acceleration during toothbrushing

On the plot, each peak represents one brush stroke, with the maximum net acceleration occurring when the brush changes direction. The value of 800 is enough to capture most strokes, since most strokes produce a change in acceleration of around 1,000. If you brush your teeth quite gently, you might need to decrease this threshold, or you could be left brushing your teeth forever, never racking up a large enough score to stop.

Next, place the first if block to detect acceleration large enough to indicate a brush stroke and, inside that, place a second if block that checks the value of count. If the value has exceeded the number you set earlier in strokes per point, then the program adds 1 to score and displays the new score value. Finally, the program checks whether score is greater than 9 and, if it is, displays a smiley-face icon.

MicroPython Code

Here is the MicroPython version of the code.

```
from microbit import *
from math import sqrt

strokes_per_point = 50
old_mg = 0
count = 0
change_in_mg = 0
score = 0
mg = 0
display.show(str(score))
```

```
while True:
    x, y, z = accelerometer.get_values()
    mg = sqrt(x*x + y*y + z*z)
    change_in_mg = mg - old_mg
    old_mg = mg
    if change_in_mg > 800:
        count += 1
        if count > strokes_per_point:
            score += 1
            display.show(str(score))
            count = 0
            if score > 9:
                display.show(Image.HAPPY)
```

The MicroPython code mirrors the Blocks code almost exactly, except that the overall strength of the force has to be calculated since MicroPython has no built-in function for this for the micro:bit.

Things to Try

You could use this project as a *pedometer*—a device that measures how many steps you take when walking or running. To do this, try simplifying the code to get rid of the score variable, because we are now only interested in the number of steps (equivalent to the strokes when toothbrushing). You'll want your code to keep track of steps and then, when you press button A, display the number of steps you've taken. To test it, tuck the project into your sock (remove the toothbrush first) and walk around while counting your steps in your head. Then see how many steps the pedometer says you've taken. If the measurement isn't accurate, you may need to change the acceleration threshold from 800 to make the pedometer more or less sensitive.

EXPERIMENT 7: LOGGING ACCELERATION TO A FILE

The Plotter built into Mu is great if you don't mind keeping your micro:bit tethered to your computer with a USB cable. However, sometimes the Mad Scientist finds it useful to record readings on the micro:bit remotely for later analysis.

In this experiment, you'll use your micro:bit to record accelerometer readings in a file saved on the micro:bit. You can wave the micro:bit around and do various other tests on it and then look at charts of the measured acceleration.

The program will take about 60 readings per second, and it can record about 45 seconds' worth of samples before the micro:bit runs out of memory.

What You'll Need

For this experiment, you'll need:

Micro:bit

3V battery pack

Construction

1. This project uses the micro:bit's local filesystem, which is not available in Blocks code at the time of this writing. That means you'll have to use Python for this experiment. Download the code from *https://github.com/simonmonk/mbms/*, along with instructions for downloading and using the book examples. The Python file for this experiment is *Experiment_07.py*. Load the program onto your micro:bit.

2. When you turn on the micro:bit, its display will show an X. This means it isn't recording any readings. When you click button A, the icon will change to a check mark, and the micro:bit will start recording. It will stop recording when you press button A again, at which point it will save the accelerometer readings in a file that you can transfer to your computer.

    To test this out, press **button A**, wave the micro:bit around, and press **button A** again.

3. To transfer the file containing the readings to your computer, use Mu's *Files* feature. Connect your micro:bit to your computer with a USB cable and click the **Files** icon in Mu's toolbar (Figure 5-9).

    The bottom of the window now has two columns. On the left are the files saved on the micro:bit. In Figure 5-9, there is only one file, *data.txt*. On the right are the files in Mu's code directory, which is in your home directory under *mu code*.

Figure 5-9: Mu's File feature in action

To copy the file from the micro:bit, just drag it from the left area to the right area in the Mu window.

As with Experiment 6, earlier in the chapter, if the Files feature doesn't work in Mu, you may have an old micro:bit that needs an update. In this case, follow the instructions at *https://support.microbit.org/support/solutions/articles/19000019131-how-to-upgrade-the-firmware-on-the-micro-bit/* to update your device.

4. Once you've transferred the data from the micro:bit to your computer, you'll plot the data by importing the file into a spreadsheet, such as Excel or Google Sheets.

    The procedure is a bit different depending on which spreadsheet software you use. I'll show you how to use Google Sheets since it's free. You just need to be logged into a Google account.

    Visit *https://docs.google.com/spreadsheets/* and click the **Blank** option in the *Start a new spreadsheet* area. Then, from the Google Sheets menu, select **File ▸ Import**. From the pop-up window that appears, select the **Upload** tab and navigate to the *data.txt* file that you copied onto your computer.

    The recorded data should appear in the first column of your spreadsheet. Select the column and click **Insert ▸ Chart** to create a chart of the data, like the one shown in Figure 5-10.

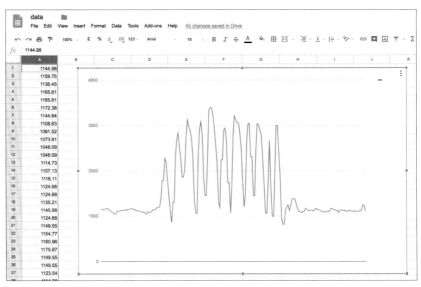

Figure 5-10: Charting data recorded on the micro:bit

Code

Here is the MicroPython code for this experiment:

```
from microbit import *
from math import sqrt
import os

filename = 'data.txt'

recording = False
display.show(Image.NO)

while True:
    if button_a.was_pressed():
        recording = not recording
        if recording:
            display.show(Image.YES)
            try:
                os.remove(filename)
            except:
                pass
            fs = open(filename, 'w')
            else:
            display.show(Image.NO)
            fs.close()
    if recording:
        x, y, z = accelerometer.get_values()
        net = sqrt(x*x + y*y + z*z)
        fs.write(str(net))
        fs.write('\n')
        sleep(10)
```

The micro:bit can save only a limited amount of data, so import the os package, which will let you delete any data already on your micro:bit.

Set the data filename as data.txt. You can change the name of this file by altering the value of the filename variable, though I recommend keeping it as is until you've gotten the program working.

Make a variable called recording to keep track of whether the project is recording or not. This is toggled between True and False in the main while loop every time button A is pressed to start and stop the recording of data. That's what the line recording = not recording does: if recording is True, the code sets it to False, and vice versa.

Create a `while True` loop to run forever. Inside this loop are two `if` statements. The first tells the micro:bit what to do when button A is pressed, and the second checks whether we are in recording mode. When button A is first pressed, recording starts, the screen displays the YES image, and the `os.remove` method deletes the existing data file.

We've put the `remove` command within a `try: except:` Python structure. This ensures that if any error occurs (in particular, if the data file can't be deleted because it's not there), the program ignores the error and doesn't crash the program.

The program then opens the file with a mode of `w`, which means you can write in it. When button A is pressed again, the screen displays the NO image and the file closes.

Also contained in the main `while` loop is another `if` block that writes a reading from the accelerometer to the file as long as `recording` is `True`. If the program runs for too long and fills up all the file space, it will give an error. However, the data that it wrote before it ran out of room will still be available.

The `sleep` command at the end of the recording slows down the recording process so that you don't run out of memory too quickly.

Things to Try

This experiment will let you measure accelerations in various practical situations. You could, for example, record the accelerations on your micro:bit when you throw it into the air. If you plan to try this, then it's wise to take a few precautions:

▶ Choose an environment with soft ground. That way, if you fail to catch your micro:bit, it might survive. During experiments like this, it's not unusual for a battery to come loose or fall out of its holder.

▶ Don't throw your micro:bit somewhere where it may hit someone in the head.

▶ Attach the micro:bit to the battery pack. Elastic bands are good for this.

▶ Put your micro:bit in a case. A Kitronik MI:pro case with MI:power battery backpack, shown in Figure 5-11, is a great choice. If you use this case, you won't need a battery pack, because the case contains a tiny 3V battery.

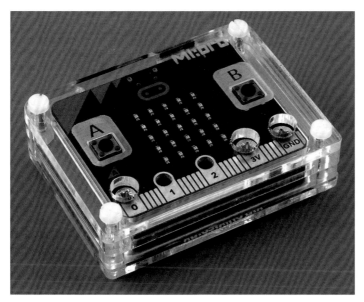

Figure 5-11: Kitronik MI:pro case with MI:power battery backpack

You could also measure acceleration by securely attaching a string to the micro:bit and swinging it gently in circles. Again, be careful as the micro:bit could easily become detached from its string, breaking or hurting someone nearby.

The filesystem used by the micro:bit is very limited; it has only about 40KB available for storage. Therefore, previous files may be erased every time you flash a new program onto the micro:bit.

PROJECT: ACCELERATION DISPLAY

Difficulty: Easy

This project, shown in Figure 5-12, allows you to see the micro:bit's acceleration on its display. When the micro:bit is at rest, the middle row of LEDs on the display will be lit. If you rapidly move the micro:bit up, then the line of LEDs will move up the display, like an elevator, in response to the increase in the net force. Similarly, if you quickly move the micro:bit down, the line will move down, indicating the reduced effect of gravity, as if you were accelerating downward in an elevator.

Figure 5-12: A Micro:bit acceleration display

What You'll Need

For this project, you just need a micro:bit. However, it's useful to have a battery pack if you want to make this project more mobile.

You could also use the MI:pro case and MI:power combination shown in Figure 5-11.

Construction

1. Go to *https://github.com/simonmonk/mbms/* to access the book's code repository and click the link for **Acceleration Display**. Once the program has opened, **Download** and then copy the hex file onto your micro:bit.

    If you prefer to use Python, then download the code from the same website. For instructions for downloading and using the book's examples, see "Downloading the Code" on page 34. The Python file for this experiment is *ch_05_Acceleration_Display.py*.

2. Try moving your micro:bit up and down, watching how acceleration in various directions affects the reading. If you can take your micro:bit for a ride in an elevator, watch as the display shows whether you're going up or down.

Code

The code first takes a reading of acceleration and then uses a bit of math to decide which row of the display to light up.

Blocks Code

Here is the Blocks code for this project.

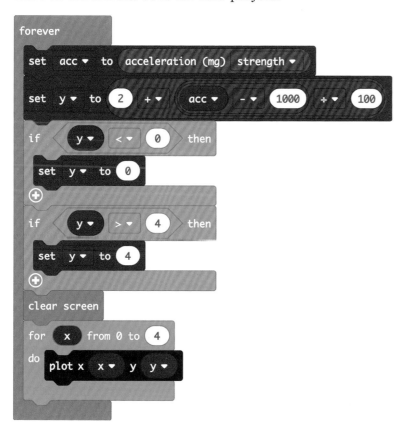

All the code for this project is contained in the forever block. It first reads the net acceleration and then calculates a value for y. Here, y represents the row of the display that will light up. When the micro:bit is stationary, the only force acting on it is gravity, at a net force of 1,000 mg (1 g). Therefore, the program subtracts 1,000 from the net acceleration and divides the result by 100, so that each 1/10 of g will cause a change of 1 row in the display. Finally, we add 2 to this result to display row 2 (the middle row of LEDs, if you start counting at 0) by default.

Use two `if` blocks to make sure that the value of y remains between 0 and 4 (for the 5 rows). To draw the correct row, first clear the screen so the old row isn't displayed along with the new reading. Then use a `for` loop to loop over each of the five LEDs for that row, which are represented by values of x, to turn them on.

MicroPython Code

Here's the MicroPython equivalent of the code:

```
from microbit import *
from math import sqrt

while True:
    x, y, z = accelerometer.get_values()
    acc = sqrt(x*x + y*y + z*z)
    y = int(2 + (acc - 1000) / 100)
    display.clear()
    if y < 0:
        y = 0
    if y > 4:
        y = 4
    for x in range(0, 5):
        display.set_pixel(x, y, 9)
```

This code is similar to the Blocks code, but you have to calculate the net acceleration yourself.

SUMMARY

The micro:bit's accelerometer opens up a lot of opportunities for projects that detect the movement or orientation of a micro:bit. In this chapter, you've explored some interesting ways you can use the accelerometer.

You've also learned how to plot data coming from the micro:bit using Mu's Plotter and record readings into a file to chart and analyze later.

6
MAD MOVEMENT

n this chapter, you'll use a few different types of motors to make two of the most impressive projects in this book: an animatronic head that swivels its eyes and talks and a robotic, remote-controlled rover. These toys can amuse the Mad Scientist for hours.

EXPERIMENT 8: MAKING A SERVOMOTOR MOVE

One way to get things moving is to use a servomotor, like the one shown in Figure 6-1.

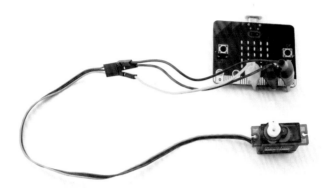

Figure 6-1: A servomotor connected to a micro:bit

A *servomotor* is a small, low-powered motor with an arm you can control with your program. Unlike most other motors, servomotors don't rotate all the way around. Instead, they have a 180-degree range of motion. Your program can set the position of the motor arm to a particular angle.

In this experiment, you'll learn how to connect a servomotor to a micro:bit and investigate how the servomotor moves.

What You'll Need

Micro:bit

Servomotor   A 9-g servomotor is ideal. A micro:bit has just enough power to drive a small servomotor but would struggle with a full-size one. Choose a servomotor that is 3V compatible. See the appendix for more details.

Alligator clip-to-male jumper cables   These connect the micro:bit to the servomotor. (You can also use male-to-male jumper cables; see below.)

USB connection to a computer, Monk Makes Power for micro:bit or USB battery pack   An AAA battery box may (depending on the servo) work, but AAA batteries may not provide enough voltage for the servomotor. See the appendix for options for powering your micro:bit.

Instead of using the alligator clip-to-male jumper cables, you could use the more common male-to-male jumper cables by pushing one end of the cable into the servomotor connector and clipping an alligator cable to the other end. However, you would need to make sure the connections don't accidentally short out. In general, alligator clip-to-male jumper cables will be a useful thing to have in your micro:bit toolbox, so it's worth getting some.

Construction

Connect the servomotor to your micro:bit.

1. Go to *https://github.com/simonmonk/mbms/* to access the book's code repository and click the link for **Experiment 8: Servomotors**. Once the program has opened, click **Download** and then copy the hex file onto your micro:bit. If you get stuck on this, head back to Chapter 1.

    If you prefer to use Python, you'll find the code for this at the same website. For instructions for downloading and using the book's examples, see "Downloading the Code" on page 34. The Python file for this experiment is *Experiment_08.py*.

2. Servomotors come with different kinds of arms that can be attached to the cog-like shaft of the motor. For

this project, select a simple arm like the one shown in Figure 6-1.

NOTE *Your servomotor should come with a little screw. This is intended to fix the arm more permanently to the shaft. If you're planning to do the animatronic head project, keep the screw nearby; otherwise, put it in a safe place for later use.*

3. Connect your micro:bit to the servomotor using the alligator clip, as shown in Figure 6-2.

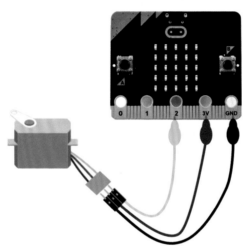

Figure 6-2: A servomotor connected to a micro:bit

4. The servomotor has three connections, which are color-coded:

> **Control**   Orange or yellow (This pin controls the position of the servomotor's arm.)
>
> **+V**   The red wire is the positive power wire. Servomotors ideally use 5V, but most small servomotors will also work with the 3V of a micro:bit.
>
> **GND**   Usually brown, sometimes black (This is the negative power wire.)

5. Once powered up, the servomotor arm should jump to a 90-degree position, perpendicular to the servomotor. The micro:bit will use this position as a reference point. If the arm isn't at 90 degrees, take it off and put it back on so that it is, as shown in Figure 6-1. If you plan to make the animatronic head, use the small screw to fix the arm in place.

6. You now have a functioning servomotor! Pressing button A should move the servo arm 10 degrees in one direction. Pressing button B should move the servo arm 10 degrees in the other direction. If you press both buttons together, the current angle of the arm should scroll across the micro:bit's display.

Code

Both programs follow the same approach of first setting the servomotor's angle to 90 degrees and then waiting for button presses to increase the angle, decrease it, or display it.

Blocks Code

Here is the Blocks code for this experiment.

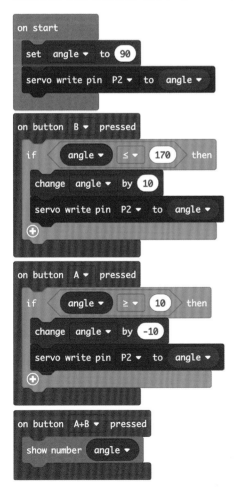

We use a variable called angle to remember the current angle of the servomotor. We define the angle variable in the on start block and give it an initial value of 90. When the next servo write pin block runs, it moves the servo arm to the position set in angle so that the arm jumps to a 90-degree angle on power-up.

If you press button A, the on button A pressed block is called. If you press button B, on button B pressed is called. The buttons work in a similar way. The block for button A first checks whether the angle is still greater than or equal to 10 and, if so, subtracts 10 from the angle to move the arm in one direction. The block for Button B checks whether the angle is less than or equal to 170 and, if it is, adds 10 to the angle and moves the arm in the opposite direction. Both use the servo write pin to set the servo to the new angle as a result of the buttons being pressed.

MicroPython Code

Here is the MicroPython version of the code.

```
from microbit import *

def set_servo_angle(pin, angle):
    duty = 26 + (angle * 51) / 90
    pin.write_analog(duty)

angle = 90
set_servo_angle(pin2, angle)

while True:
    if button_a.was_pressed() and angle >= 10:
        angle -= 10
        set_servo_angle(pin2, angle)
    if button_b.was_pressed() and angle <= 170:
        angle += 10
        set_servo_angle(pin2, angle)

if button_a.is_pressed() and button_b.is_pressed():
    display.scroll(str(angle))
```

The MicroPython code works in much the same way as the Blocks code. But unlike in the Blocks code, there is no predefined function to set the servomotor to a particular angle. Fortunately, we can write our own method using a little math and PWM analog outputs (see "Making an Analog Signal: Pulse Width

Modulation" on page 8), which generate the pulses that our servomotor expects (more on this next).

How It Works: Servomotors and Pulses

You control servomotors by sending them a series of repeating electrical pulses—in this case, from the micro:bit. The pulses are generated by turning a pin on and off very quickly. The servomotor knows how to act based on how long the pulse has been high (3V) over a certain time period. The proportion of time that the signal is high, and thus the pin is on, is known as the *duty cycle*. This is different from the actual amount of time that the signal is high, which is known as the *pulse width*. We refer to this amount of time, measured in milliseconds (ms), as a *width* because we can visualize pulses as a square wave (see Figure 6-3). The total length of the wave (between each *on* pulse) is the *period*.

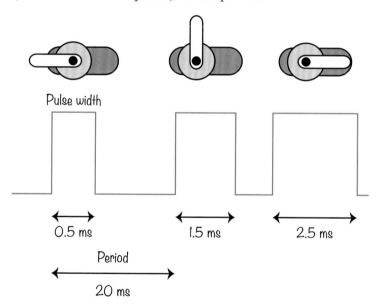

Figure 6-3: Pulses controlling a servomotor

A servomotor expects to receive a pulse every 20 milliseconds, or at a rate of 50 pulses a second. Each pulse has a width between 0.5 milliseconds and 2.5 milliseconds.

As you can see in Figure 6-3, the length of the pulse determines the position of the arm. If the pulse width is 0.5 ms, the servomotor's arm will be at one end of its range (0 degrees). If the pulse width is 1.5 ms, the arm will be at its center position (90 degrees). And if the width is 2.5 ms, the arm will be at 180 degrees.

Conveniently, the micro:bit's default frequency for analog output (see Chapter 1) is 50Hz—exactly the right frequency for a servomotor. However, even the longest pulse is only 2.5 ms out of a total period of 20 ms, meaning that the longest pulse will only be high about one-tenth of the time.

Figure 6-4 helps to explain how the numbers used in the MicroPython set_servo_angle function were calculated. The *duty value* is the number supplied to the set_servo_angle function. This value must be between 0 and 1023, where 0 is no pulse at all and 1023 is a pulse so long that it lasts until the next cycle.

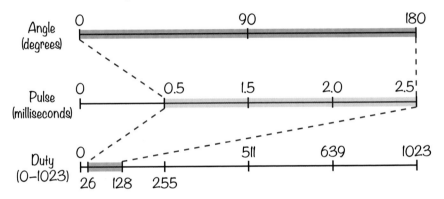

Figure 6-4: Correspondence of duty values to pulse widths to servo arm angles

The top line of Figure 6-4 shows the servo arm's range of positions in terms of angles from 0 to 180 degrees. The code must convert this into a different range of numbers, from 0.5 to 2.5, that are the pulse widths corresponding to those angles.

To convert an angle in degrees to a pulse width in milliseconds, we'll start with the information that a pulse of length 0.5 ms is equal to an angle of 0 degrees and a pulse of length 2.5 ms is equal to an angle of 180 degrees. We can then find the degrees per millisecond by dividing the range of degrees (180) by the range of pulse lengths (2), giving 90 degrees per millisecond. Then, to calculate the pulse length of a new angle, we begin with our baseline, 0.5 ms and add the angle we are using divided by 90, the degrees per millisecond.

Let's consider an example. If the angle is 0 degrees, the pulse length in milliseconds will be 0.5 (0.5 + 0/90 = 0.5). If the angle is 90 degrees, the pulse length will be 0.5 + (90/90) = 1.5 ms. And if the angle is 180 degrees, the pulse length will be 0.5 + (180/90) = 2.5 ms.

Now we have a formula for the pulse length for an angle:

$$pulse\_length = 0.5 + angle/90$$

But, referring to Figure 6-5 again, we need to convert the pulse length in milliseconds to a duty value between 0 and 1023 because the set_servo_angle function expects a value in that range.

The duty value (0 to 1023) is calculated by multiplying the pulse length in milliseconds by the number of steps per millisecond (1023/20 ≈ 51). For example, a pulse length of 1.5 milliseconds would require a duty value of 1.5 × 51 ≈ 77.

In other words:

$$duty\_value = pulse\_length \times 51$$

Combining these two formulas, we have:

$$duty\_value = (0.5 + angle/90) \times 51$$

This can also be written as (with rounding):

$$duty\_value = 26 + angle \times 51/90$$

So the values used in the write_analog function are between 26 and 128. This range reflects the fact that the pulses are quite short in comparison to the maximum duty cycle value of 1023.

PROJECT: ANIMATRONIC HEAD (MIKE THE MICRO:BIT ROBOT)

Difficulty: Hard

This animatronic head, shown in Figure 6-5, makes a great project for a Halloween display. A servomotor moves a pair of ping-pong ball eyes from left to right, and it uses the micro:bit's display as a mouth. When the head talks, the lights simulate a simple animation of lips moving.

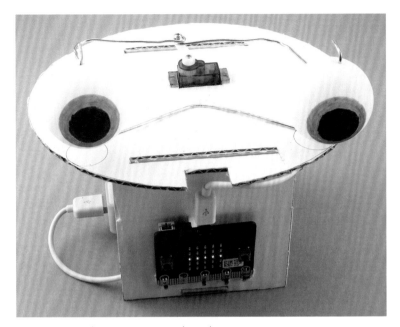

Figure 6-5: The animatronic head

What You'll Need

For this project, you'll need the following items:

Micro:bit

Servomotor A small 9g servomotor is ideal.

Amplified speaker For the head's voice (Monk Makes Speaker for micro:bit is a good choice.)

3 × Alligator clip-to-male jumper cables To attach the servomotor

3 × Alligator clip cables To attach the speaker

2 × Ping-pong balls For the eyeballs (Balls without any writing or logos on them make decoration easier. Grab a couple of spares, just in case.)

Thick card Roughly letter or A4 card that is 3 mm or more thick

Paper and access to a printer To print a template for cutting out the framework

2 × 3-inch (75 mm) paper clips These are used to make the axles that allow the eyes to swivel and the frame that connects the eyes to the servomotor. Ideally, these are 1.5 mm in diameter.

Adhesive tape   For sticking the cardboard together and sticking various things to the cardboard

Blu-Tack adhesive putty   To attach the speaker to the cardboard

Paper glue   To stick the template onto the cardboard

Scissors and/or craft knife

Paint or pens   To draw the eyeballs

A drill with 5/64-inch (#47) or 2-mm bit   To make holes in the ping-pong balls

Pliers   To bend the wire

Construction

We'll begin by creating the eyeballs.

1. Grab two ping-pong balls (Figure 6-6) and prepare to decorate. The best way to draw an eyeball is to first find two small circular items, one a bit bigger than the other (perhaps the lid of a toothpaste tube and a ring) and then, using the circular items and a pencil, trace two concentric circles on the surface of the ping-pong ball. Once this is done, color the outer ring (the iris) and the inner circle (the pupil) two different colors.

Figure 6-6: Decorating an eyeball

2. Once you've drawn on the two balls to make them look like eyes, drill three holes in each, through which you'll feed wires. Using a pencil, make three marks: with the pupil of the eye facing you, make one mark at the bottom of the ball, one at the top, and a final mark on the back opposite the pupil. The holes on the top and bottom will be used to thread the eyeball onto a vertical wire to hold it in position. The hole in the back will be used to attach the second wire, which will move the eye left and right.

WARNING   *Using a drill can be dangerous! For this part of the project, a responsible adult should use the drill or at least supervise. The main precaution is to keep the ping-pong ball on a flat surface and hold it from the sides. Then drill from above, as shown in Figure 6-7. If you have a vise to clamp the ball in place, that's even better.*

Figure 6-7 Drilling the eyeball

3. Once you're happy with the position of the marks, grab a drill bit with a diameter slightly larger than that of the paper clip wire and drill into the ping-pong ball. For the 1.5 mm paper clips, a 5/64-inch (#47) or 2-mm drill bit is perfect.

4. Make a frame for the eyeballs. Start by completely straightening out a paper clip. Then make the three bends labeled A, B, and C in Figure 6-8.

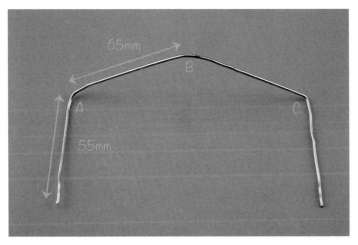

Figure 6-8: Where to bend the paper clip to make the eye frame

5. Make bends A and C first. Both should be 90 degrees. These will hold the eyeballs upright. Next, make a slight bend at the center of the wire—this is bend B. Bend B will provide something to tape onto the cardboard chassis you will make next. Place the eyeballs on the frame as shown in Figure 6-9.

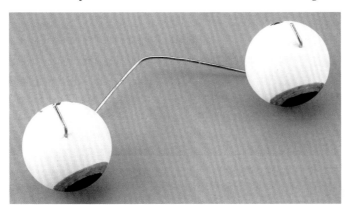

Figure 6-9: Placing the eyes on the frame

6. Once you've mounted the eyeballs, you can bend the 10 mm tips of the wires over the eyeballs to prevent them from falling off, as shown in Figure 6-9. Use pliers to avoid hurting your fingers.

NOTE   *These bends don't have to be perfect the first time. Plan to make some adjustments as you build the project.*

7. Now let's start assembling the cardboard chassis that will hold everything together. To make this easier, I've provided a template that can be downloaded with the code for the book. The template is in the folder *other downloads* and is called *Animatronic_Head_Template*. It is available in PDF, PNG, and SVG formats. Download it, print it out, and fix it onto a slightly larger piece of cardboard, as shown in Figure 6-10.

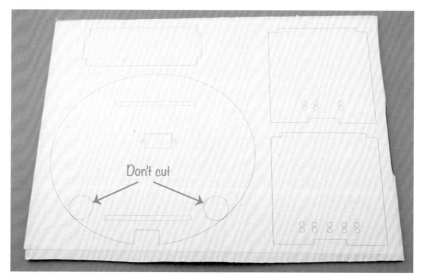

Figure 6-10: The paper template glued to a piece of cardboard

8. Except for the places noted in Figure 6-10, cut along the lines. This will give you the pieces of cardboard shown in Figure 6-11. For many of the cuts, scissors will be fine, but

you'll really need a craft knife to make the cutouts for the slots and servomotor.

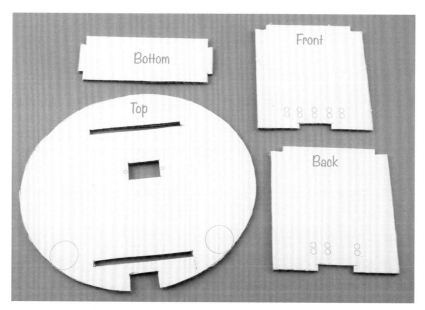

Figure 6-11: The cardboard pieces for the chassis

WARNING *Only use a craft knife with adult supervision. Craft knives are very sharp, and it's easy to accidentally cut yourself with them.*

9. Attach the micro:bit to the front card by poking holes through the pairs of small circles marked on the front of the board. Then, for each pair of holes, clip an alligator clip through the back. These clips will both allow you to make electrical connections and keep the micro:bit securely attached to the card.

    Attach the alligator clips in the order suggested in Figure 6-12: red to 3V, black to GND, and yellow to pin 0. Make sure to attach two alligator clips each to the GND and 3V connections of the micro:bit.

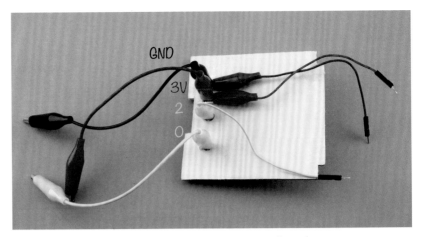

Figure 6-12: Connecting cables to the micro:bit

10. To connect the micro:bit to the servomotor, you'll need to attach three alligator clips with male jumper cables on one end. Clip a yellow alligator clip with a male jumper cables to pin 2. Then, pull back the insulating sheath on the red alligator clip already attached to the 3V, expose some of the metal, and clip the red alligator clip with male jumper cable to the red clip. Do the same with the black GND cable, pulling back the insulating sheath of GND's black alligator clip and attaching the black alligator to male jumper cable (Figure 6-13).

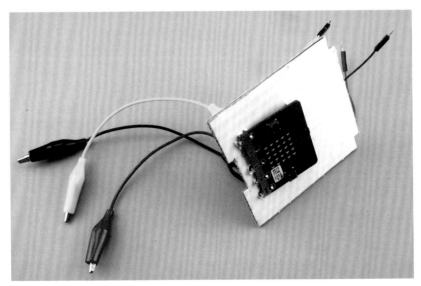

Figure 6-13: Attaching the micro:bit to the front of the cardboard

11. Now slot the bottom piece of cardboard into the front piece and fix it in place using adhesive tape (Figure 6-14).

Figure 6-14: Attaching the bottom card to the front card

12. Now attach the speaker to the back card; use adhesive tape or putty to secure it. Clip the other end of the three alligator clips you've connected to the micro:bit (black, red, and yellow) to their corresponding ports on the speaker (GND, 3V, and 0, respectively). Affix the back card to the bottom card using adhesive tape. As you can see in Figure 6-15, your chassis should now be in a U shape. The red, blue, and yellow alligator clips here are waiting to be attached to the servomotor.

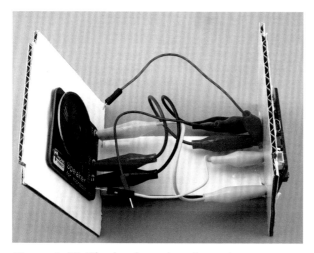

Figure 6-15: The back card and speaker

13. With the servomotor arm fixed in place at 90 degrees (review the end of Experiment 8 if needed), push the servomotor through the top card (the round face), threading the servomotor's wire through first. Notice that the servomotor's shaft is not in the center of the servomotor but rather toward one end; that end should be at the end of the servomotor cutout that is closest to the center of the top card (see Figure 6-16).

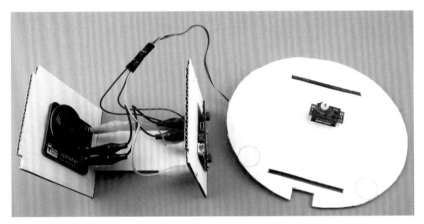

Figure 6-16: Connecting the servomotor

14. Attach the male jumper pins to the servomotor's socket, as described in Experiment 8.

15. With everything connected together, attach the top piece of cardboard to the rest of the chassis. The two circles marking the position of the eyes on the top piece should face the micro:bit end of the chassis (refer to Figure 6-5).

16. Now attach the eyeballs and wire frame to the top card, as shown in Figure 6-17. Adjust the bend in the middle of the wire frame so that the eyeballs are centered over the two circles drawn on the top card. Then use adhesive tape to hold the wire in place. Give the eyeballs a little spin to make sure they can turn freely.

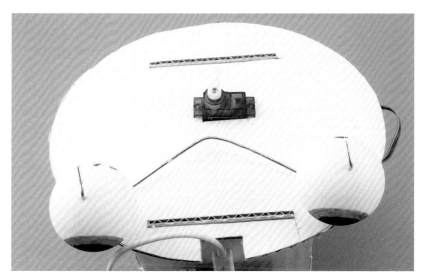

Figure 6-17: Connecting the eyeballs

17. To connect the servomotor to the eyeballs, you'll need to straighten out the other paper clip, as shown in Figure 6-18.

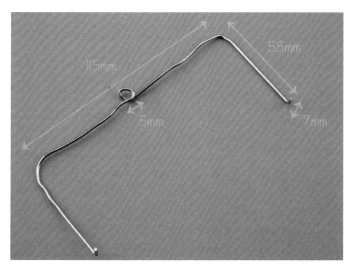

Figure 6-18: Making the connector between the servomotor and the eyeballs

18. To make the loop in the center of the wire, wrap the wire tightly around a small screwdriver shaft (shown in Figure 6-19). The diameter of the screwdriver I used was about 3 mm, making the outside diameter of the little loop around 5 mm. Use a screwdriver with a sharp end that's narrower than the shaft, or it will be hard to slide the wire off once it's been bent. Make sure to straighten the legs of the wire as shown.

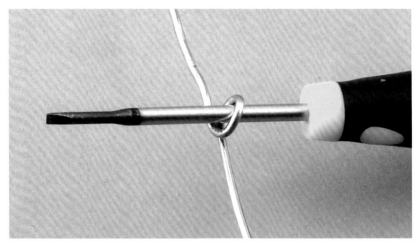

Figure 6-19: Using a screwdriver to bend a loop in the servomotor-to-eyeball connector

NOTE   *Screwdrivers are sharp and it takes strong hands to bend the thick wire around the screwdriver, so you might need adult help for this bit.*

19. Adjust the connector by bending the paper clip until the arms are the same distance apart as the holes in the backs of the eyeballs. Then, hook the wire into the backs of the eyeballs. Use one of the screws provided with the servomotor to fasten the loop to the tip of the servomotor's arm, as shown in Figure 6-20. Depending on how snugly the servomotor fits its cutout in the top card, you may need to tape down the motor to prevent it from moving.

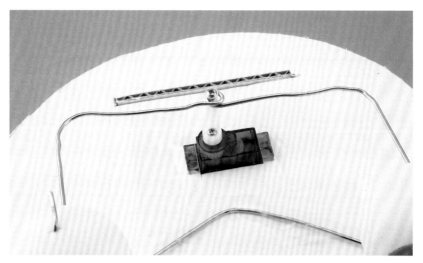

Figure 6-20: Fitting the servomotor–eyeball connector to the card

Now connect the USB cable from your micro:bit and try the project out!

Code

First, we'll use the code from Experiment 8 to test that our mechanisms are all working. Once you're sure everything is working well, switch to the code for this project.

Using the Code from Experiment 8 as a Test

Go to *https://github.com/simonmonk/mbms/* and click the link for **Experiment 8: Servomotors**. Once the program has opened, click **Download** and then copy the hex file onto your micro:bit. If you prefer to use Python while testing, download the Python file, *Experiment_08.py*, from the same website.

Once the code is loaded, try pressing the A and B buttons to move the servomotor's arm left and right. You will probably need to adjust the connector wire slightly so that the eyes are lined up and move freely left to right. Only take the servomotor arm a few steps to either side of the 90-degree position or the linkage might get jammed.

Using the Real Project Code

Once you're satisfied that the eyeballs are moving the way they should, you can switch to the real project code.

Because this project uses a speech library, it is only available in MicroPython form. Go to *https://github.com/simonmonk/mbms/* to download the Python file is *ch_06 _Animatronic_Head.py*.

Flash it onto your micro:bit and then gently poke the robot. The accelerometer should pick up the movement and tell your animatronic head to swivel its eyes, say something witty, and then look straight ahead again. For a video of this project, go to my YouTube channel (*https://www.youtube.com/watch?v=FAJTS2Z8ZDA*).

The software for this project does two things: it detects when the head is poked, and it triggers talking events after random periods of time have elapsed. The code is rather long, so instead of showing all of it, I'll just highlight the key parts. If you want to follow along, load the code into Mu.

To make it seem as though your animatronic head has a mind of its own, we use the Python random library to give us random numbers, which will trigger random events:

```
import random
```

Rather than give the head a free range of eye movements, we keep a list of possible eye angles in the array eye_angles. When we want to set the eyes in a random direction, we simply use the random function from the random library to take an angle from this array.

If you've played around with the head, you've probably noticed that it doesn't know that many sentences. The phrases it speaks are contained in the array sentences, shown here:

```
sentences = [
"Hello my name is Mike",
"What is your name",
"I am looking at you",
"Exterminate exterminate exterminate",
"Number Five is alive",
"I cant do that Dave",
"daisee daisee give me your answer do"
]
```

As with the eye angles, when we want a sentence, we'll take it from this array. As you can see in the code, with robotic speakers, sometimes it's better to spell a word phonetically than spell it correctly.

The three lip images for the speech animation are created as custom Image objects. Here is the one for the resting lip animation, a horizontal line:

```
lips0 = Image("00000:"
              "00000:"
              "99999:"
              "00000:"
              "00000")
```

Each of the five rows in Image is a string representing one row of the display; each digit in the string represents the brightness from 0 to 9 of a particular LED.

The images for the lips are held as an array in the variable lips:

```
lips = [lips0, lips1, lips2]
```

Both the speaking and the lip animation are controlled by the function speak, which takes in the sentence to speak as a parameter:

```
def speak(sentence):
    words = sentence.split()
    for i in range(0, len(words)):
        display.show(random.choice(lips))
        speech.say(words[i])
    display.show(lips0)
```

To make sure that the lips animate as the words are spoken, we use the split method to break the sentence into the list of individual words saved as words. Then, for every word in the list, we display one of the lip images (chosen at random using the choice method from the random library) and have speech speak the word. When all the words have been spoken, the lip display shows the default lips0 image.

Next, we define an act() function:

```
def act():
    set_servo_angle(pin2, random.choice(eye_angles))
    sleep(300)
    speak(random.choice(sentences))
    set_servo_angle(pin2, 90)
    sleep(2000)
```

The act function does three things: it moves the eyes at random by setting the servomotor to a random angle, it selects a sentence to speak by calling speak, and it resets the eyes by setting the servomotor angle back to 90 degrees. To allow for some time between each step, the code makes a call to sleep.

Here is the main body of our code that uses all of our variables and functions:

```
while True:
    new_z = abs(accelerometer.get_z())
    if abs(new_z - base_z) > 20:
        base_z = new_z
        act()
    if random.randint(0, 1000) == 0:
        act()
    sleep(50)
```

In the main body of the code, we have a while True loop, which means the commands execute until the code is signaled to stop. This is useful for when you need code to respond continuously to input. Here, we want to be ready for the sudden change in acceleration caused by a tap on the robot's head. First, the loop registers the acceleration from the accelerometer. Then, it uses the abs function to get the magnitude of the acceleration—in this case, we don't care about the direction, just how large the acceleration is.

In the first if statement, we check whether the acceleration value has changed by more than 20 mg (milligravities). If so, the base acceleration is updated to the new acceleration (ensuring that the next time through the loop, the acceleration has to change by *another* 20 mg), and act is called.

In the second if statement, we give the animatronic head a bit of randomness. The code picks a random number between 0 and 1,000. If it is equal to 0 (a 1 in 1,001 chance), the act function is called. Even though this probability is very low, since the value gets checked hundreds of times a second, the head springs into action several times a minute.

Things to Try

Try using a USB battery or AAA battery box to power the head instead of keeping it tethered to your computer with a USB cable.

If you want to change up your head's speech, go into the code and add more sentences to the sentences array.

The speech library produces rather quiet speech that is also quite indistinct. You can improve this a little by connecting a bigger amplified speaker.

If you want to add a bit more to the project, take a look at the code examples here: *https://microbit-micropython .readthedocs.io/en/latest/tutorials/speech.html*. In this code, the speech library is used to produce singing.

PROJECT: ROBOT ROVER

Difficulty: Hard

In this project, we'll create a robotic rover. Using a clever app called Bitty Controller, you'll be able to control the little buggy with your Android phone (Figure 6-21). The Mad Scientist likes to use the rover to deliver notes to the lab assistants.

WARNING   *We're going to use some low-cost chassis kits for this project, but the wires that come with these kits are usually loose—meaning you'll need to solder the wires onto the motors. This is the only project in the book that requires soldering. The soldering isn't difficult, but it is dangerous and you can easily get burned. So please find an adult to do this part.*

Figure 6-21: A micro:bit-controlled roving robot

What You'll Need

For this project, you'll need the following items:

Micro:bit

Android phone

Kitronik Motor Driver Board for micro:bit (V2)   To control the motors

Low-cost robot chassis kit   Includes two gear motors and a 4 × AA battery box

4 × AA batteries

Bitty Controller App for Android   From Google Play Store (about $5)

Assorted screwdrivers   Suitable for both the nuts and bolts on the chassis and the screw terminals on the motor controller board

Soldering equipment   To attach the wires to the gearmotors

Blu-Tack adhesive putty   To attach the motor control board and micro:bit to the chassis

If you search eBay or Amazon for robot chassis, you'll find low-cost robot chassis kits like the one shown in Figure 6-22. Look for one that includes a 4 × AA battery box and two gear-motors (motors with a built-in gearbox).

Figure 6-22: The low-cost robot chassis used by this Mad Scientist

Construction

Maybe the trickiest part of this project is the chassis kit assembly. Getting the screws and bolts in the right place will require precision and finesse.

Not all chassis will be the same, so instead of taking you through the assembly step-by-step, I'll just give you some high-level advice. Your chassis should come with instructions, although they may be somewhat cryptic. In general, you'll need to attach the gear motors, the castor wheel (the wheel that can turn freely in any direction), the motor controller, and the micro:bit.

1. Solder the supplied wires onto the motor terminals, as shown in Figure 6-23a–c. If you work quickly, a good soldering joint can be made by melting a generous

amount of solder onto the motor terminal and then pressing the wire onto the dome of solder with the iron (Figure 6-23a).

a. Soldering the first motor wire

b. The first wire soldered in place

c. Both wires soldered to the gearmotor

Figure 6-23: Soldering wires onto the motor

2. It doesn't matter much which terminal you attach the red wire to. Just be sure to be consistent between the two motors. That is, if you decide to attach the red wire to the right-hand terminal on one motor, solder the red wire to the right-hand terminal of the other motor as well.

3. Remove the layer of paper covering the chassis. Figure 6-24a–e shows the assembly process at a high level. You may find that your chassis is different.

a. Attaching the first motor

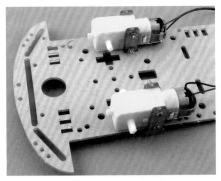

b. Both motors fitted

c. Spacers bolted onto the castor

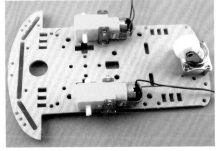

d. The castor bolted to the chassis

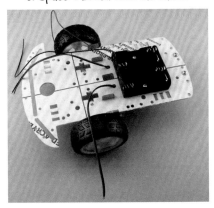

e. The completed chassis with
battery box bolted to the top

Figure 6-24: Building the chassis

Here are a few things to remember when building the chassis:

▶ Don't overtighten the nuts and bolts, as doing so can cause
the plastic chassis to crack.

▶ When attaching the gearmotors, put the nuts toward the
inside of the chassis so that the motor-fixing bolts that
stick out beyond the nut won't get in the way of the wheels.

▶ If the bolt doesn't turn smoothly into the nut, try turning the bolt counterclockwise a little to find the start of the nut's thread. You may want to use a pair of small pliers to grip the nut while you turn the screw with a screwdriver.

4. Once you have the chassis securely assembled, attach the micro:bit to the Kitronik Motor Driver board with the micro:bit's screen facing outward, as shown in Figure 6-25. Line up the micro:bit's edge connector carefully with the socket on the Motor Controller and press it firmly into place. When your controller looks like Figure 6-25, stick it to the chassis using adhesive putty, also shown in Figure 6-25.

Figure 6-25: Attaching the micro:bit

5. It's time to wire up the motors and battery box. First, look at Figure 6-26 to see what you are aiming for. Unscrew the screw on the relevant terminal, place the wire firmly inside the terminal, and then screw the screw back in fairly tightly.

Figure 6-26: Wiring up

6. Make the following connections:

▶ Red (positive) wire from the battery box to the screw terminal marked RED + on the Motor Controller

▶ Black (negative) wire from the battery box to the screw terminal marked BLACK - on the Motor Controller

▶ Red wire from the left (as viewed from the back of the rover) motor to the screw terminal on the Motor Controller labeled MOTOR1 P12

▶ Black wire from the left motor to the screw terminal on the Motor Controller labeled MOTOR1 P8

▶ Red wire from the right motor to the screw terminal on the Motor Controller labeled MOTOR2 P0

▶ Black wire from the right motor to the screw terminal on the Motor Controller labeled MOTOR1 P16

7. In this project, you'll use an Android app to control the rover via Bluetooth. You can find the code that runs on the micro:bit at *http://www.bittysoftware.com/downloads.html#controller*. Click the link **micro:bit hex file for**

Kitronik Buggy - no pairing required and download the hex file. This file is also available with the book downloads in the *Other Downloads* folder. Next, connect the micro:bit to your computer with a USB and copy the downloaded hex file onto your micro:bit. You won't need batteries just yet.

8. To install the app on your Android phone, open Google Play and search for *Bitty Controller*. You'll have to pay a few dollars for the app. Download and install the app.

9. We're one step away from trying out the project! Put four AA batteries into the battery box. These will power both the motors and your micro:bit, meaning you can disconnect the USB cable—it's time to set your rover free.

Open the Bitty Controller app (Figure 6-27) and click **Scan**. This should find your micro:bit. In Figure 6-27, it's called *BBC micro:bit [gaviv]*. Click this and the RC-style controller of Figure 6-28 should appear. The micro:bit's display should also show a C, indicating that it's connected to your phone.

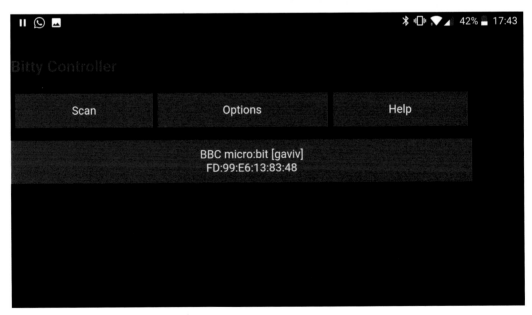

Figure 6-27: Starting Bitty Controller

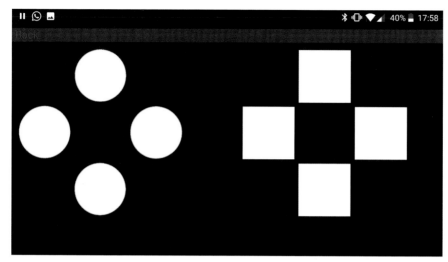

Figure 6-28: The Dual D-Pad Controller

Use the top and bottom square buttons to control the forward and backward motion of the rover. To turn the rover left and right, use the left and right round buttons.

For a first test, do something simple: flip the rover onto its back and use the app to control the wheels. Do they spin? Once the rover passes that test, put it on the floor and try driving it around. If you don't like the controller layout, head to Options on the Bitty Controller app to find other layouts.

The rover may move forward when you tell it to go backward and vice versa. If this happens, your wires are swapped: switch the red and black wires for motor 1 and motor 2. If your rover drives around in a circle, swap one of the pairs of wires on one of the motors.

When you want to turn off your rover (a good way to make your batteries last longer), just lift one end of one of the batteries out of the battery holder. Ta-da! Now you have a crude switch.

How It Works: Motors and the Flow of Electricity

The direction of the gear motors is controlled by the direction of the current flowing through them—if you reverse the direction, you reverse the motors. As you can see in Figure 6-29, a motor turns clockwise if connection A is positive and connection B is negative. If A and B are reversed, so that A is negative and B is positive, the motor moves the other way.

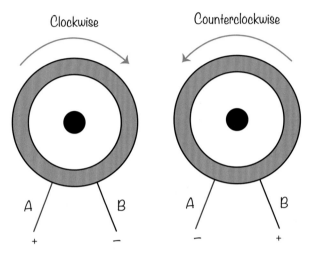

Figure 6-29: Controlling the direction of a motor

The Kitronik Motor Controller contains a chip that controls the direction of current in two motors. It also supplies the relatively high current that the motors need.

SUMMARY

We covered a lot of ground in this chapter. First, we learned about servomotors and how to set one up using the micro:bit. Then, we built two complex projects: the animatronic head and the remote-controlled rover. Along the way, we learned about PWM and current flow and picked up a few basic programming tricks to boot.

Now that you know how to make things move, you can start thinking about other projects you'd like to make. What things would you like to have move all on their own? In the next chapter, we'll look at how to deal with time.

7

TIME TRAVEL

kay, so we won't actually build a time machine in this chapter, but you'll measure time by making a binary clock and a clock that can speak. We'll also conduct an experiment to test how well your micro:bit can keep time. All the experiments and projects in this chapter use MicroPython only.

EXPERIMENT 9: KEEPING TIME

The aim of this experiment is to make a micro:bit clock that keeps good time. That means programming the micro:bit to tick at precisely one-second intervals.

One way to do this would be to use the sleep function, as in the code shown here. Note that this code is not a full program, so don't try to run it. The sleep command stops the micro:bit from doing anything for however long you specify. In our case, the delay is 1,000 milliseconds (1 second).

```
seconds = 0

while True:
    sleep(1000)
    seconds += 1
```

In this example, after each 1 second delay, the program adds 1 to the seconds variable, which counts the number of seconds that have passed. This loop repeats indefinitely and, as a way of marking time, works for a bit.

The problem is that the clock will gradually fall behind because we haven't accounted for the time it takes the micro:bit to add 1 to the seconds variable. In this example, adding 1 to the variable won't take much time at all, but if the program got any longer—for example, by telling the clock to also display the time or even speak the time, which we'll try later—then the delay could become significant. It would also be unpredictable, as the time lost may not be the same every time the program loops.

Therefore, a better way to keep time is to use the running _time function. This function returns the number of milliseconds that have passed since the micro:bit was last reset, and it's not affected by how long other parts of your code take to do things.

In this experiment, we'll use the running_time function to calculate just how slow or fast our micro:bit clock runs.

What You'll Need

To carry out this experiment, you just need two things:

Micro:bit
USB cable

Construction

1. Find the code at *https://github.com/simonmonk/mbms/*. The Python file for this experiment is *Experiment_09.py*. Open the program in Mu and flash it onto your micro:bit.

2. Once you've successfully programmed the micro:bit, press button **B**. You should find that the micro:bit sets the seconds to 0 and starts counting up from there.

3. Set a timer on your phone or another device for exactly 16 minutes 40 seconds. Start the timer and, at the same time, press button **B** to reset the micro:bit's second count. At the end of the timed period, press button **A** to freeze the clock and make a note of the number of seconds displayed.

Because this experiment is all about timing, it's important to start the timer at exactly the same time that button B is pressed and to press button A as soon as the timer sounds.

This will be easier if a friend helps: one of you can operate the timer, while the other operates the micro:bit.

The reason for setting the timer to 16 minutes 40 seconds is that this is 1,000 seconds. If the micro:bit's second count is greater than 1,000, then the clock is running fast, and if the count is less than 1,000, the clock is running slow. My micro:bit's second count was 989, indicating that the micro:bit's internal clock was running about 11 parts in 1,000 too slow.

Make a note of your micro:bit's second count. You'll use it in the projects in this chapter to make your clock more accurate.

Code

Here is the MicroPython code for Experiment 9:

```
from microbit import *

seconds = 0
last_time = 0

while True:
    now = running_time()
    elapsed_ms = now - last_time
    if elapsed_ms >= 1000:
        seconds += 1
        last_time = now
    if button_a.was_pressed():
        display.scroll(str(seconds))
    if button_b.was_pressed():
        seconds = 0
        display.show("0")
        sleep(100)
        display.clear()
```

The program uses two variables:

last_time   Keeps track of the last time that the clock ticked

seconds   Keeps track of the number of seconds that have passed since the micro:bit was last reset and the clock started running

I find it useful to think of the clock *ticking*, like a regular clock. That is, it does something at regular intervals.

The main while loop uses the running_time function to find out how long the micro:bit has been running in milliseconds. It stores that number in a variable called now. It then calculates how many milliseconds have elapsed since the last tick by subtracting last_time from now.

If the number of milliseconds elapsed is greater than or equal to 1,000—in other words, greater than or equal to 1 second—then the seconds variable increases by 1. Then we reset the number of milliseconds elapsed to 0 so we can count elapsed time over again.

We use two if statements to program button A and button B. If you press button A, the micro:bit will display seconds, or the time that's passed since the program started running. If you press button B, the seconds count will reset to 0.

How It Works: Keeping Time

The micro:bit's processor uses a crystal oscillator (an electronic component used to keep time accurately) that should be accurate to better than 30 parts per million. However, for my micro:bit, it was inaccurate by 11,000 parts per million for some reason.

To get a truly accurate clock, you'd need to use a dedicated RTC (Real Time Clock) chip and separate crystal oscillator. At the time of writing, no RTC chips are available specifically for the micro:bit. Although they can be made to work with the micro:bit, this is a fairly tricky process. Therefore, it's probably best not to rely too much on either of the clocks you'll build in this chapter, but these projects are fun and will teach you important skills.

PROJECT: BINARY CLOCK

Difficulty: Easy

In this project, shown in Figure 7-1, you'll create a clock that shows you the time in *binary*. Binary is a numbering system used in computers. You can learn more about it in the "How It Works: Telling the Time in Binary" on page 176. A binary clock displays hours, minutes, and seconds as separate binary numbers on the micro:bit's LED display.

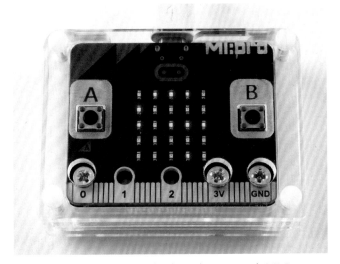

Figure 7-1: The binary clock in the Kitronik MI:Pro case

Figure 7-2 shows the binary numbering system on the micro:bit. It might seem confusing at first, like a random pattern of LEDs, but I'll explain how it works shortly. Plus, the Mad Scientist loves to show off their skill at reading binary clocks to impress their friends!

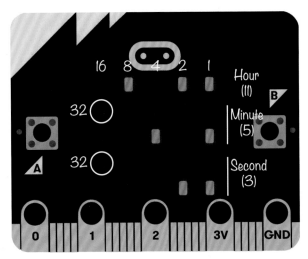

Figure 7-2: Reading the binary clock

How to Read the Binary Clock

Our binary clock is a 24-hour clock. To read the clock, start at the top row of LEDs, shown in Figure 7-2. This row represents the hours. Each column of LEDs stands for a number. From right to left, the columns represent the numbers 1, 2, 4, 8, and 16. By adding these five numbers, you can create every possible value between 1 and 24. To work out the hour, add the column numbers for the LEDs that are lit. In the case of Figure 7-2, that's 1, 2, and 8, which add up to 11. So, the hour is 11.

The next two rows represent the number of minutes, and the bottom two rows represent the number of seconds. Both minutes and seconds are indicated by a row with the same 1, 2, 4, 8, and 16 LEDs. However, since we need to be able to count all the way to 60 for minutes and seconds, these values are indicated by an additional LED (worth 32) on the preceding row. As shown in Figure 7-2, the 32 LED for minutes is the leftmost LED on the second row from the top, and the 32 LED for seconds is the leftmost LED on the fourth row from the top.

In Figure 7-2, the 4 and 1 minute LEDs are lit, indicating 5 minutes. The 2 and 1 second LEDs are lit, indicating 3 seconds. All together, this display is saying the time is 11:05:03.

Holding down button A will make the numbers on the clock advance very quickly, allowing you to set the time. You just need to be ready to stop the clock as soon as it reaches the correct time.

You can see a video of the clock in operation, including how the time is set, at *https://www.youtube.com/watch?v=v26gYo5OG0g*.

What You'll Need

For this project, all you need is your micro:bit and a power source.

If you plan to keep your clock running constantly, then you should use a USB power adapter or other long-term power source for the micro:bit (see the appendix) to save on batteries. You might also want to get a case for the micro:bit to make your clock look nicer.

Remember that the clock won't keep perfect time, so you'll need to reset it fairly often.

Construction

1. The code for this project is in MicroPython, because the math required would be extremely tricky to do in the Blocks code. Download the code from *https://github.com/simonmonk/mbms/*. The file for this project is *ch_07_Binary_Clock.py*.

2. Before loading the program onto your micro:bit, open it in Mu and change the current time so it's accurate. You should probably set it to a few minutes before the current time, just to be safe. You can adjust it later by holding down button A.

    Change the time by altering the following lines. Remember that this is a 24-hour clock, so, for example, 6:00 PM is 12 + 6 or 18 hours.

```
hours = 8
minutes = 25
```

You should also change the value of the adjust variable to the number of parts per thousand by which your micro:bit's clock is slow or fast. My micro:bit ran slow by 11 seconds in 1,000, so I set adjust to –11 to speed it up a little (note the negative sign). If the micro:bit had run fast by, say, 10 seconds per 1,000, then I would have set adjust to 10 to slow things down a tiny amount.

Code

This project involves quite a lot of math. We'll also use some more advanced programming features, such as *two-dimensional arrays*, which are much easier to implement in Python than in Blocks code.

Let's break the code into chunks, starting with the lines that assign the correct binary values to each LED:

```
# hhhhh
# m
# mmmmm
# s
# sssss
```

```
sec_leds = [[4, 4], [3, 4], [2, 4], [1, 4], [0, 4], [0, 3]]
min_leds = [[4, 2], [3, 2], [2, 2], [1, 2], [0, 2], [0, 1]]
hour_leds = [[4, 0], [3, 0], [2, 0], [1, 0], [0, 0]]
adjust = -11
```

The first four lines of code aren't actually part of the program. They're *comment lines*, or notes, reminding you which LEDs display the hours, minutes, and seconds. These can serve as a useful reference when setting the LED coordinates in the arrays that follow.

The next three lines are the two-dimensional arrays mentioned earlier. These assign the proper LED coordinates to the seconds, minutes, and hours. Remember that arrays are like variables, except they hold multiple elements. The arrays we're using here are called *two dimensional* because their elements are also arrays. For example, the first element in the first array, sec_leds, is [4, 4]. This identifies the first LED used to display the number of seconds as the LED with an x coordinate of 4 and a y coordinate of 4. That's the LED in the bottom right corner of the display. Figure 7-3 shows the coordinates of the individual LEDs that make up the display.

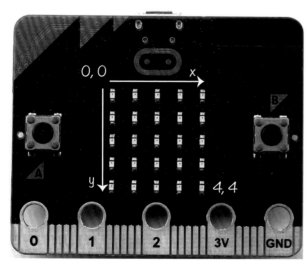

Figure 7-3: LED display coordinates

As you can see, the top left LED has the coordinates [0, 0] and the bottom right [4, 4].

Next, we have the hours, minutes, and seconds variables to keep the current time:

```
hours = 8
minutes = 25
seconds = 1
adjust = -11
```

We'll use the adjust variable to correct the clock's speed. See "Construction" on page 167 if you're not sure how to do this.

Here is the function that turns the LEDs on or off to indicate a value in binary:

```
def display_binary(value, num_bits, leds):
    v = value
    for i in range(0, num_bits):
        v_bit = v % 2
        display.set_pixel(leds[i][0], leds[i][1], int(v_bit * 9))
        v = int(v / 2)
```

We have three separate binary numbers (hours, minutes, and seconds) to display, and the display_binary function works for all of them. It takes a number to display (value), the number of LEDs to use in displaying the number (num_bits), and an array of LEDs to use (leds). It uses these three values to display the three parts of the time—the seconds, minutes, and hours—on the micro:bit.

To keep track of the time, you use two variables:

```
last_time = 0
tick = 1000 + adjust
```

The variable `last_time` records the last time that the clock ticked, and the variable `tick` holds the duration of the clock's tick in milliseconds. The default value for `tick` is `1,000` + `adjust`, but this value will change when you press button A to set the time.

Here is the code to update the time:

```
def update_time():
    global hours, minutes, seconds
    seconds += 1
    if seconds > 59:
        seconds = 0
        minutes += 1
        if minutes > 59:
            minutes = 0
            hours += 1
            if hours > 23:
                hours = 0
```

The function `update_time` adds 1 to the number of seconds every time it is called. When the `seconds` count gets past 59, it resets to 0 and increases the number of `minutes`. It does the same thing for the `hours`. We use nested `if` statements to accomplish this.

This is the code to display the hours, minutes, and seconds in binary:

```
def display_time():
    display_binary(seconds, 6, sec_leds)
    display_binary(minutes, 6, min_leds)
    display_binary(hours, 5, hour_leds)
```

We put this code inside the `display_time` function, which calls the `display_binary` function defined earlier.

Here is the main `while` loop that makes the clock run quickly when someone presses button A and normally otherwise. It also contains the code to keep time.

```
while True:
    if button_a.is_pressed():
        tick = 10
    else:
        tick = 1000 + adjust
    now = running_time()
    elapsed_ms = now - last_time
```

```
if elapsed_ms >= tick:
    update_time()
    display_time()
```

The first part of the loop checks whether button A is being pressed. If it is, the code reduces tick to 10 milliseconds. Otherwise, it sets it to 1000 + adjust.

Finally, we write the code that allows the clock to keep time. The function running_time returns the number of milliseconds since you last reset your micro:bit. Each time the program loops, we calculate how much time has elapsed since the clock last ticked. The loop does the following:

1. Gets the current running_time and stores it in a variable called now

2. Calculates the value for elapsed_ms by figuring out the difference between the variables now and last_time

3. Updates the time if elapsed_time is greater than our tick time of 1 second

4. Sets last_time to now, resetting the millisecond count to zero

How It Works: Telling the Time in Binary

Using this binary clock to tell the time is a little tricky. It's especially hard to figure out the seconds, which are likely to have changed before the code has calculated them. But there's a reason the binary system exists.

Most of us are familiar with the decimal system of writing numbers. *Decimal* is the Latin word for 10, and in the decimal system, we use 10 different symbols (the digits 0 through 9). If we need to write a number greater than 9—say 15—then we use two digits. Because of the position of the 1 in the number 15, we know that it actually represents the number 10.

The following table shows the numbers from 0 to 10 in binary. Note that in decimal, we don't write *leading zeros*. We wouldn't write 15 as 0015, for example. In binary, however, it's customary to write numbers with the leading zeros to give the numbers the same number of digits. That's just the way computer scientists roll. So in this case, all the binary numbers are four digits long.

Decimal	Binary
0	0000
1	0001
2	0010
3	0011
4	0100
5	0101
6	0110
7	0111
8	1000
9	1001
10	1010

In theory, computers could store numbers like this by using 10 different voltages to represent the digits 0 to 9, but they don't. Instead, computers use a system called binary. Rather than having 10 possible values for a digit, a binary digit (also called a *bit*) can represent either a 0 or a 1. Computers use binary because the *transistors* they are made with are really good at being either off (0) or on (1). That means they have only 2 possible states, which is much easier than giving them 10 possible states. Also, the math behind binary logic means that a computer can do reliable arithmetic on binary numbers far more easily than it were dealing with numbers in decimal.

Just as in the familiar decimal system, the binary system combines digits to represent bigger numbers. Whereas each digit position in a decimal number increases by a factor of 10—going from 1, to 10, to 100—each binary digit increases by a factor of 2—from 1, to 2, to 4, and so on. For example, the four-digit binary number 1010 has 1s in the 16- and 2-digit positions and 0s in the other positions. In decimal, it's 16 + 2, or 18.

It turns out that you don't need very many binary digits to represent some pretty big numbers. For example, eight binary digits put together (called a *byte*) can represent a decimal number anywhere between 0 and 255. Make that 16 bits, and the number goes up to 65,535. A computer with 64 bits is able to do everything using those binary 64 digits, and it can represent a number between 0 and 18,446,744,073,709,551,615. Incidentally, a micro:bit has a

32-bit processor, able to represent numbers between 0 and a very respectable 4,294,967,295. The MicroPython function running_time that we've been using uses a 32-bit number. This means that it will not run out of numbers for 4,294,967,295 ÷ 1,000 ÷ 60 ÷ 60 ÷ 24 = 49.7 days.

PROJECT: TALKING CLOCK

Difficulty: Easy

Sometimes the Mad Scientist is so busy with test tubes and chemicals and alarming plumes of smoke that they can't check the time. And then they forget to eat! That's when a talking clock comes in handy. This project (Figure 7-4) announces the time every hour, or whenever you press button A.

Figure 7-4: A talking clock

You can see this project in action at *https://www.youtube .com/watch?v=iNjXEK8RUtU*.

What You'll Need

For this project, you'll need the following items:

Micro:bit

3 × Alligator clip cables   To connect the micro:bit to the speaker

Speaker for micro:bit   To play the sounds (Use a Monk Makes Speaker or see Chapter 2 for other speaker options.)

Power adapter   See the appendix for options on keeping your micro:bit powered without batteries.

If you're planning to keep your clock running, then use a USB power adapter or another long-term power solution to save on batteries.

You may want to build a case for the clock or attach the micro:bit and speaker to the same piece of cardboard you used for the light-controlled guitar project from Chapter 3.

Construction

1. The code for this project is in MicroPython, as the speech library is not currently available in the Blocks code. Download the code from *https://github.com/simonmonk/mbms/*. The file for this project is *ch_07_Talking_Clock.py*.

2. Before loading the program onto your micro:bit, open the file in Mu and change the hours and minutes to your current time. Also change adjust to the amount by which you want to adjust your clock. See "Construction" on page 167 of the binary clock project if you're not sure how to do this.

3. Connect a speaker to the micro:bit with one alligator clip on pin 0 and the other alligator clip on the pin marked IN on the speaker. Use the other two clips to provide power to the speaker, as shown in Figure 7-4.

4. Wait for the time you set in step 2 to arrive. Then connect the micro:bit to a power source. Note that for this project, there is no other mechanism to set the time.

Code

For clarity, we'll work through this code in sections.

Here's the code that does the time keeping—in other words, that makes sure the hours and minutes are correct. It's almost the same as the time-keeping code from the previous project. The main difference is that, instead of showing the time on the LED display, the code will speak the current time.

```
digits = ["no", "1", "2", "3", "4", "5", "6", "7", "8", "9",
"ten", "eleven", "twelve", "thirteen", "fourteen", "fifteen",
"sixteen", "seventeen", "eighteen", "nienteen"]
tens = ["no", "no", "twenty", "thirty", "forty", "fifty"]
```

```
preamble = "The time is "
am = "aye em"
pm = "pee em"
```

We use several variables and arrays to hold a set of words that the micro:bit will speak.

The speech library contains recordings of some common words. It can say single digits like 1, 2, or 3, but for numbers 10 and over, it pronounces each digit separately, which is not what we want our talking clock to do. That's why we have to spell out numbers greater than 10. Notice that *nineteen* is misspelled as *nienteen* to make it sound right when the synthesized voice says it. The array called `digits` holds the text for each number up to 19. The clock should never speak the 0 digit, so we just set it to the word no.

The tens array does a similar job with numbers that are multiples of 10. We already accounted for all the numbers up to 19 with the digits array, so we don't need to worry about the first two elements in the `tens` array, which will never be spoken. We set these to no as well.

The `preamble` variable contains the text that the micro:bit will speak before it announces each time. The am and pm variables contain phonetic versions of the AM/PM indicator. The micro:bit will speak one of these after reading the time.

Here's the code with the function that actually speaks the time. Appropriately enough, it's called `speak_the_time`.

```
def speak_the_time():
    h = hours
    am_pm = am
    if h >= 12:
        am_pm = pm
    if h > 12:
        h = h - 12
    if minutes == 0:
        # The time is twelve pm exactly
        speech.say(preamble + digits[h] + " "
                        + am_pm + " exactly")
    else:
        if minutes < 10:
            # The time is twelve o four pm
            speech.say(preamble + digits[h] + " o "
                    + digits[minutes] + " " + am_pm)
```

```
    elif minutes < 20:
        # The time is twelve eighteen pm
        speech.say(preamble + digits[h] + " "
            + digits[minutes] + " " + am_pm)
    else:
        mins_tens = int(minutes / 10)
        mins_units = minutes % 10
        if mins_units == 0:
            # The time is twelve twenty pm
            speech.say(preamble + digits[h] + " "
                + tens[mins_tens] + " " + am_pm)
        else:
            # The time is twelve twenty four pm
            speech.say(preamble + digits[h] + " "
    + tens[mins_tens] + " " + digits[mins_units] + " " + am_pm)
```

This function is fairly complex, as it has to account for our different ways of expressing the time.

This clock speaks in the 12-hour format but stores the hours in 24-hour format, so the first thing speak_the_time does is decide whether the time is AM or PM. It subtracts 12 from the hour variable once hour reaches 13.

Next, the nested if statements cover the following possible cases:

▶ If the time is exactly on the hour, say something like The time is twelve pm exactly.

▶ Otherwise, if the minutes are less than 10, add an o, to say something like The time is twelve o four pm.

▶ For two-digit minutes less than 20, use the digits array and say something like The time is twelve eighteen pm.

▶ For other two-digit minutes of 20 or over that are multiples of 10, use the tens array to say something like The time is twelve twenty pm.

▶ And, where the minutes are not multiples of 10, say something like The time is twelve twenty four pm.

Last comes the main while loop:

```
while True:
    if button_b.is_pressed():
        speak_the_time()
    now = running_time()
```

```
elapsed_ms = now - last_time
if elapsed_ms >= tick:
    elapsed_seconds = int(elapsed_ms / tick)
    update_time(elapsed_seconds)
    blink()
    last_time = now
```

This loop checks whether button B has been pressed or an hour has passed and speaks the time if either event has occurred. It also calls a function called `blink`. That flashes the heart icon on the screen to reassure you that the clock is working, even if it is silent most of the time.

How It Works: Teaching the Micro:bit to Speak

The MicroPython speech library opens up all sorts of possibilities in your projects, as you saw back in Chapter 6. The sound quality isn't perfect, but it does add loads of fun to your projects.

The speech library itself is based on the concept of *phonemes*: building blocks of sound. When you use the `say` function, the text to be spoken is first translated into a series of phonemes. Because of the strangeness of spoken language, this often doesn't work perfectly—hence the misspelling of *nineteen* in the code for this project to help the `say` function pronounce the word more accurately.

You can read much more about this speech library at *https://microbit-micropython.readthedocs.io/en/latest/tutorials/speech.html*.

SUMMARY

Hopefully you now have a good sense of how to make a clock using a micro:bit and use it to display or literally tell you the time.

In the next chapter, the Mad Scientist turns their attention to psychological experiments.

8
MAD SCIENTIST
MIND GAMES

n this chapter, the Mad Scientist turns their attention to the source of their genius—the mind! First, you'll learn about your nervous system by testing how long it takes to react to stimuli. You'll then build a lie detector that measures galvanic skin resistance, one of the factors used in polygraph tests. Use it on your friends. If they get warm and sweaty when you question them, they might be lying to you!

EXPERIMENT 10: HOW FAST ARE YOUR NERVES?

Difficulty: Medium

People process stimuli with nerve cells, or neurons. *Nerves* are the wiring of the human body, carrying signals from the brain to the rest of the body and back. When you touch a hot stove, you seem to feel the pain right away. But compared to the copper wires in most electronics, nerves actually work pretty slowly.

By measuring the time it takes your brain to respond to a signal, you can estimate how fast that signal travels through your nerves. In this experiment, you'll press button A with either your hand or your foot whenever the micro:bit screen goes blank, and the micro:bit will measure the time it takes you to react.

We'll make a cardboard pedal to go around the micro:bit so you can press the button without hiding the display (Figure 8-1).

Figure 8-1: Turning a micro:bit into a foot-operated switch

What You'll Need

For this project, you'll need:

Micro:bit

Piece of cardboard about 8 inches × 4 inches (20 cm × 10 cm)   Thick, strong cardboard works best. It doesn't need to be exactly these dimensions.

Craft knife   To cut and score the cardboard

Blu-Tack adhesive putty   To attach the micro:bit to the cardboard

Construction

We'll start by creating a cardboard pedal that will fold in half around the micro:bit.

1. Draw two parallel lines roughly halfway down the cardboard and about 1/4 inches (5 mm) apart, as shown in Figure 8-2. You'll fold the cardboard along these lines. Also, mark the rectangular part of the cardboard you want to cut out. Make sure the hole you'll make is big enough for the micro:bit's screen.

2. Draw your knife along the lines for the hinges, cutting only about halfway through the cardboard. This is called *scoring* the cardboard. Be careful not to cut all the way through. Then cut out the rectangle you marked.

Figure 8-2: Making the hinge and cutting out a place for the micro:bit display

WARNING   *Using a craft knife can be dangerous! For this part of the project, a responsible adult should use the craft knife or at least supervise.*

3. Place the micro:bit inside the folded cardboard so you can see its screen. Stick small pieces of Blu-Tack to the center of the micro:bit and then stick the micro:bit to the cardboard. When you gently press the top of the hinge, you should be able to feel button A click (see Figure 8-1). Check that when the switch is pressed, it doesn't also accidentally press the Reset button on the back of the micro:bit. If it does, try using larger pieces of Blu-Tack.

4. This experiment uses MicroPython only. Find the code at *https://github.com/simonmonk/mbms/* The Python file for this experiment is *Experiment_10.py.* Open the file in Mu and load it onto the micro:bit.

Testing Your Nervous System

The program that controls this experiment uses Mu's REPL to report results and give you instructions. (Revisit Chapter 1 if you need a refresher on how the REPL works.) Because it uses the REPL, this is one experiment where you'll need to keep your micro:bit connected to your computer using the USB cable. We'll use the micro:bit's display and button A to measure reaction times.

Open Mu's REPL. The instructions in the REPL area of the window will ask you to hold down button A whenever the micro:bit's display shows a cross and release it as fast as you can when the display goes blank.

Here is what you might see in the REPL on a typical run. The instructions are broken into sections for clarity.

```
TEST 1 - USING your hand
Hold the switch down while the cross is showing.
Release momentarily when the display blanks.
Repeat 5 times.
Press ENTER when ready to start the test
```

To start the first test run, hold down the switch with your dominant hand and press ENTER once on your computer. A cross should appear on the screen. After a random delay of between 3 and 7 seconds, the display will go blank. When it does, release the button as fast as you can. The REPL should show you the number of milliseconds between when the display went blank and when you released the switch. For example:

```
252
```

The cross should then light for another random period, so hold down the switch to try the test again. The experiment should repeat a total of five times.

If you let go of the switch *before* the display has blanked, the REPL should display the message You let go too soon. This will record a time of 0, invalidating the experiment, and you'll have to restart the whole process. If this happens, reset the micro:bit by unplugging its USB cable and plugging it back in.

After you've completed the five tests, the micro:bit's display should remain blank, and the REPL will show your individual reaction times as well as the average of all your times. For example:

```
252
264
264
282
The average time using your hand was 262.7999 ms
```

You should then be prompted to repeat the experiment using your foot. Move the micro:bit pedal onto the floor. This time, position your foot over the micro:bit and gently use it to press the switch. Getting this right may take a little practice.

Now repeat the test for your foot. Once you've captured five reaction times, the REPL should display the average, and you'll be prompted to enter the following two measurements, as in this example:

```
The average time using your foot was 368.3999 ms
Enter the distance from the back of your neck to your fingers in
cm: 107
Enter the distance from the back of your neck to your toes in cm:
188
```

The program will use these measurements to calculate how long it took your brain to realize that the display had blanked in milliseconds—this is the thinking time. And it will calculate the speed at which the signal traveled to your hand or foot in meters per second—this is the transmission speed.

```
Thinking time (ms): 123.3036
Transmission speed (m/s): 13.03703
```

"How It Works: Measuring Your Reaction Time" on page 191 will explain how these calculations are made and just how meaningful they are (or aren't).

Code

We use a variable n to hold the number of reaction time readings to take for each test. In this case, five readings are taken.

```
from microbit import *
import random
```

```
n = 5

def run_full_test():
    print("TEST 1 - USING your hand")
    t_hand = run_test()
    print("The average time using your hand was " + str(t_hand)
        + " ms")
    print("Now repeat the test for your foot")
    t_foot = run_test()
    print("The average time using your foot was " + str(t_foot)
        + " ms")
❶  d_hand = int(input("Enter the distance from the back of
                        your neck to your fingers in cm: "))
❷  d_foot = int(input("Enter the distance from the back of
                        your neck to your toes in cm: "))
    thinking_time = (d_foot * t_hand - d_hand * t_foot)
                / (d_foot - d_hand)
    transmission_speed = 10 * (t_foot - thinking_time) / d_foot
    print("Thinking time (ms): " + str(thinking_time))
    print("Transmission speed (m/s): " + str(transmission_speed))
```

The function run_full_test runs both the hand and foot test in turn. (We'll define the code for both those tests in a bit.) The last four lines of the code are the calculations for the thinking time and transmission speed, explained in "How It Works: Measuring Your Reaction Time" on page 191.

This function also prompts you to enter the neck-to-hand and neck-to-foot distances using the input function. The program will be able to use that information because it stores the input in variables. We store the input for the neck-to-hand distance in d_hand ❶ and the input for the neck-to-foot distance in d_foot ❷.

The input function displays the string of text that you supplied as its parameter and then waits for you to enter some text. It returns whatever you type as a string, and the int function converts that string into an integer so we can use it in the calculation.

Here's the code for the run_test function:

```
def run_test():
    print("Hold the switch down while cross is showing and
            release momentarily when the display blanks")
    print("Repeat " + str(n) + " times.")
```

```
    input("Press ENTER when ready to start the test")
    total = 0
    for i in range(0, n):
        t = get_reaction_time()
        if t > 10:
            print(t)
            total += t
        else:
            print("You let go too soon")
    return total / n
```

The run_test function runs a single test, providing the necessary instructions and collecting the required number of reaction time readings by calling get_reaction_time. If the reaction time is less than 10 milliseconds, which would mean superhuman reflexes, the program tells you that you let go too soon. Otherwise, it displays your reaction time.

```
def get_reaction_time():
    display.show(Image.NO)
    sleep(random.randint(3000, 7000))
    display.clear()
    t0 = running_time()
    while pin5.read_digital() == False: # Button A down
        pass
    t1 = running_time()
    t = t1 - t0
    return t

run_full_test()
```

The get_reaction_time function displays the cross image and then sleeps for a random period between 3 and 7 seconds. It then clears the screen and sets the variable t0 to the current running time so that the program knows when to start timing the reaction test.

When you release button A, the program exits the while loop and records the time at which it stops in t1. We calculate the reaction time by subtracting t0 from t1.

You might be wondering why the code checks for button A being pressed using pin5.read_digital rather than button_a .is_pressed. The answer is that the is_pressed function does not operate as quickly as read_digital and so would add some unwanted extra time to the reaction time measurement. We use `pin5` because the micro:bit's wiring connects pin 5 directly to button A. The micro:bit actually has a whole load of other pins than the pins 0 to 3 that we use with alligator clips. However, in this book we concentrate on the pins that are easy to access.

Things to Try

To perform further investigations into the human nervous system, try comparing different people's results or conducting the test at different times of day.

How It Works: Measuring Your Reaction Time

When you see the cross vanish, two things happen in your body, each of which takes a certain amount of time:

1. Your eyes and brain notice the change and decide to act on it.
2. A signal passes from your brain, through your nerves, to the muscles that control your hand or foot, and the hand or foot comes off the switch.

For this experiment, we assumed that the thinking time, or the time it takes you to register the change, is the same whether you're moving your hand or your foot.

We also assume that the nerve signal speed between brain and hand and brain and foot is the same. In reality, neither of these assumptions is exactly right, but the Mad Scientist can sort out those issues some other time.

Given those assumptions, the speed that the signal travels through your nerves is equal to the total reaction time minus the thinking time, divided by the distance the signal has to travel.

Taking the two distance measurements allows us to crudely calculate both the thinking time and the signal speed. Here are the variables we'll use in the calculations:

t_hand The total reaction time (thinking + acting) for the hand

t_foot The total reaction time for the foot

d_hand The distance the signal has to travel along the nerve from the brain to the hand

d_foot The brain-to-foot distance

`thinking_time`   The time it takes to register the event and start the message to the hand or foot

`transmission_speed`   The speed of the signal through the brain to the hand or foot

Now for the math. For the hand, we can use this equation:

$$\texttt{transmission\_speed} = \frac{\texttt{d\_hand}}{(\texttt{t\_hand} - \texttt{thinking\_time})}$$

Similarly, for the foot, we can use this:

$$\texttt{transmission\_speed} = \frac{\texttt{d\_foot}}{(\texttt{t\_foot} - \texttt{thinking\_time})}$$

This means that we also know the following:

$$\frac{\texttt{d\_hand}}{(\texttt{t\_hand} - \texttt{thinking\_time})} = \frac{\texttt{d\_foot}}{(\texttt{t\_foot} - \texttt{thinking\_time})}$$

Now we can use algebra to rearrange things so that we can calculate the thinking time and transmission speed. We can multiply both sides by this:

$$(\texttt{t\_hand} - \texttt{thinking\_time})$$

And we get this:

$$\texttt{d\_hand} = \frac{\texttt{d\_foot}\,(\texttt{t\_hand} - \texttt{thinking\_time})}{(\texttt{t\_foot} - \texttt{thinking\_time})}$$

If we multiply both sides by the following:

$$(\texttt{t\_foot} - \texttt{thinking\_time})$$

we get this:

$$\texttt{d\_hand}\,(\texttt{t\_foot} - \texttt{thinking\_time}) = \texttt{d\_foot}\,(\texttt{t\_hand} - \texttt{thinking\_time})$$

Multiplying across the parentheses gives this:

$$\texttt{d\_hand} \times \texttt{t\_foot} - \texttt{d\_hand} \times \texttt{thinking\_time}$$
$$= \texttt{d\_foot} \times \texttt{t\_hand} - \texttt{d\_foot} \times \texttt{thinking\_time}$$

So:

$$\texttt{d\_hand} \times \texttt{t\_foot} - \texttt{d\_foot} \times \texttt{t\_hand}$$
$$= \texttt{d\_hand} \times \texttt{thinking\_time} - \texttt{d\_foot} \times \texttt{thinking\_time}$$

$$d\_hand \times t\_foot - d\_foot \times t\_hand$$
$$= thinking\_time\,(d\_hand - d\_foot)$$

Finally, we can calculate the thinking time as the following:

$$thinking\_time = \frac{(d\_hand \times t\_foot - d\_foot \times t\_hand)}{(d\_hand - d\_foot)}$$

Now that we have the thinking time, we can use it to calculate the speed the message travels along the nerves, like this:

$$transmission\_speed = \frac{(t\_foot - thinking\_time)}{d\_foot}$$

This result will be in centimeters per millisecond. To convert the value into meters per second, we multiply by 1,000 and divide by 100. In other words, multiply by 10, and you've got the transmission speed.

PROJECT: LIE DETECTOR

Difficulty: Medium

The Mad Scientist doesn't like surprise parties, and they want to figure out whether their lab assistants are planning one for their birthday. They'll have to use their lie detector (Figure 8-3) to figure it out!

Figure 8-3: Lie detector

This project measures *galvanic skin resistance (GSR),* which is the resistance of the skin to the flow of electricity. To measure GSR, we'll use a micro:bit, and a pair of metal spoons.

What You'll Need

For this project, you'll need the following:

Micro:bit

2 × Alligator clip cables To connect the spoons to the micro:bit

2 × Spoons

You'll use the spoons to make contact with the palms of your hands. The alligator clips will attach to the spoons' handles and might scratch them, so don't use your best spoons!

Construction

1. Go to *https://github.com/simonmonk/mbms/* to access the book's code repository and click the link for **Lie Detector.** Once the program has opened, click **Download** and then copy the hex file onto your micro:bit. If you get stuck, head back to Chapter 1 for a refresher on loading programs onto your micro:bit.

    If you prefer to use Python, then download the code from the same website. For instructions for downloading and using the book's examples, see "Downloading the Code" on page 34. The Python file for this experiment is *ch_08_Lie_Detector.py.*

2. Use one alligator clip to connect the handle of one of the spoons to pin 2 and the other alligator clip to connect the other spoon to GND, as shown in Figure 8-3.

    To use the lie detector, you need two people: an operator to ask the tricky questions and a subject to answer them. The subject needs to hold the spoons so that the curved side makes full contact with the palms of their hands.

    When the operator presses button A, the dot on the micro:bit's display should move to the center of the LED. If skin resistance falls because the subject gets sweaty, the dot will move up the screen. If skin resistance rises, the dot will move down.

If the dot seems to be stuck at the top or bottom, the operator needs to press button A to center the display. The readings will gradually drift, even without any difficult questions.

After asking a question, the operator should wait three or four seconds before checking the reading on the display.

Code

This project has both Blocks and Python code. The full explanation follows the Blocks code, so if you skip to the Python code now, make sure to return to read the explanation.

Blocks Code

Here is the Blocks code for this project.

In the on start block, we place the block set pull pin, which enables the built-in 12kΩ pullup resistor. (The pullup resistor is disabled by default.)

Also in the on start block, we take a reading from pin 2, connected to one of the spoons, and store the result in the baseline variable. The program will compare any new readings against this baseline to see how much the subject's skin resistance has changed.

We then use the forever loop to take a new analog reading from pin 2 that represents your subject's GSR after answering a question you've asked. The loop subtracts the baseline reading from the new reading and divides the result by 10 to reduce it by about the right amount.

It uses this number to determine which of the five LEDs should light up. The number must fall within the range of −2 to 2 or an error will occur. Once this is done (using a couple of if blocks), the display is cleared, and the middle-column LED on the row 2 + change is lit using the plot block.

Pressing button A resets the baseline.

MicroPython Code

The MicroPython version of the code is shown here:

```
from microbit import *

pin2.set_pull(pin2.PULL_UP)
baseline = pin2.read_analog()
```

```
while True:
    if button_a.was_pressed():
        baseline = pin2.read_analog()
    reading = pin2.read_analog()
    change = int((reading - baseline) / 10)
    if (change > 2):
        change = 2
    if (change < -2):
        change = -2
    display.clear()
    display.set_pixel(2, 2 + change, 9)
```

This works in much the same way as the Blocks code, with just a few differences. Here, we use the int function to convert the change in the reading values into an integer.

We then use the set_pixel method to tell the micro:bit display which LED to light up. This method is different from the plot block because it not only asks for the x and y coordinates but also requires you to provide a brightness level between 0 and 9. We choose 9 to make it as bright and visible as possible.

How It Works: Detecting Lies Through Sweat, Voltage, and Resistance

Skin resistance is one of the measurements used by polygraph lie detectors so often seen in movies.

If you flush or start to sweat, as you might when under pressure, your skin's resistance to electricity changes. But testing for GSR might not always work. The questions you ask and how you ask them can influence a person's reactions or make them feel nervous, and GSRs will differ from person to person. Conversely, people who are more detached from their feelings can learn how to defeat the polygraph. But for the Mad Scientist's investigation into a possible surprise party, this lie detector is just the ticket.

This project converts readings of GSR into a voltage. Most metals have low resistance because metal conducts electricity well, whereas something like plastic has a high resistance because electricity does not generally flow well through plastic. We measure resistance using a micro:bit's pin, which works as an analog input (see Chapter 1) and an arrangement of two resistors called a *voltage divider* (Figure 8-4).

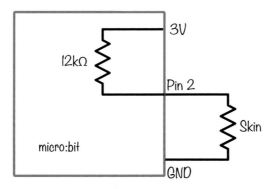

Figure 8-4: Schematic of a voltage divider

The unit of resistance is the *ohm* (Ω), and we normally abbreviate 1,000Ω to 1kΩ. In Figure 8-4, the 12kΩ resistor connected to pin 2 is a resistor built into the micro:bit's processor. This resistor pulls the voltage at pin 2 toward 3V (think of the resistors in this diagram as springs); if you weren't holding onto the spoons, this pull would be the only force acting on the pin, and the pin would register a maximum reading of 1023, or 3V.

However, as soon as the subject grasps the spoons, another resistor—the subject's skin—comes into play. This counters the 12kΩ resistor by pulling the voltage at pin 2 back toward 0V. If your skin resistance was exactly 12kΩ, then the two resistances would be pulling equally, and the voltage at pin 2 would be 1.5V.

If the subject sweats, their skin resistance falls, pulling the voltage lower. Conversely, when they recover, their skin resistance increases, allowing the 12kΩ resistor to pull the voltage up.

SUMMARY

In this chapter, we conducted a few experiments on the human nervous system. The experiment and project we completed may have dubious scientific rigor, but hopefully you enjoyed yourself along the way.

In the next chapter, we'll take measurements of the environment. Specifically, we'll work with light and temperature, which are important to the Mad Scientist's comfort.

9

ENVIRONMENTAL MADNESS

he Mad Scientist loves to measure things. In this chapter, we'll follow in the scientist's footsteps by measuring the temperature. We'll devise a temperature and light logger that will provide valuable insights into our environmental experiments. Then we'll create a plant-watering project that uses a small pump to water a houseplant automatically when its soil starts to dry out.

EXPERIMENT 11: MEASURING TEMPERATURE

Difficulty: Easy

Owning no clothes besides lab coats and having skimped on the insulation budget for the Secret Laboratory, the Mad Scientist decided to build a custom heating system. To do this, they needed to know the exact temperature in every room of the lab. Unfortunately, they soon found the measurements taken with their micro:bit were not exact enough.

The micro:bit has a function named temperature in both the Blocks and MicroPython code that returns a temperature reading in degrees Celsius. However, the sensor itself is built into the micro:bit's processor and thus is actually reporting the temperature of the micro:bit's *chip*, not the temperature of the micro:bit's surroundings.

At normal room temperature of around 20 degrees C (68 degrees F), if the micro:bit hasn't been running for too long, the sensor gives a fairly accurate reading. However, we can't be sure whether the temperature reading is correct if the micro:bit's processor has been busy and started warming up.

In this experiment, you'll investigate the difference between the temperature readings of a busy micro:bit and an idle micro:bit.

What You'll Need

To do this experiment, you'll just need two things:

Micro:bit

USB cable

You may also want a separate thermometer to check your readings against. The only other thing you'll need is some patience, as you'll need to leave this program running for half an hour to get good readings.

Construction

1. This project uses Mu's Plotter feature, which you'll need Python for, so there's no Blocks code for this. Find the code at *https://github.com/simonmonk/mbms/*. The Python file for this experiment is *Experiment_11.py*. Flash the program onto your micro:bit.. This code will take temperature readings every 20 seconds. It will also initiate micro:bit activity every 10 minutes and then idle for 10 minutes, letting you see the difference in temperature between when the micro:bit is busy and when it's idle.

2. Open both the REPL and Plotter views of Mu and press **Reset** on the micro:bit to start the reading process. A new temperature reading, and an indication as to whether the micro:bit has been busy or idle, are logged to the Plotter and REPL every 20 seconds. The busy/idle status of the micro:bit flips every 10 minutes. Figure 9-1 shows the result of logging the data for 40 minutes.

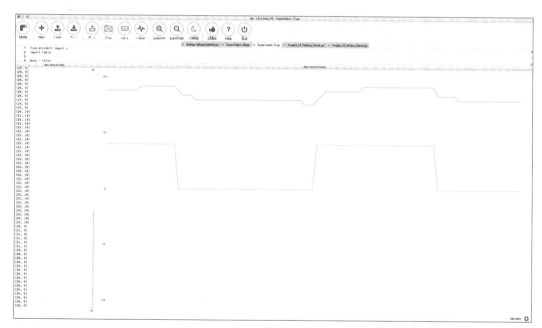

Figure 9-1: Plotting temperature readings and processor activity

The blue line shows the temperature reported by the temperature function, and the green line indicates whether the micro:bit is busy or idle. When busy, the micro:bit turns on the display and its radio interface, and it displays the message Busy over and over.

As you can see from Figure 9-1, when the micro:bit is in busy mode, the reported temperature rises by about 3 degrees C. When it goes back to idle mode, the temperature drops. Note that the temperature in the room, as measured by a separate thermometer, remained at 20.0 degrees C throughout the experiment.

Code

The MicroPython code for the experiment needs to do two things: flip the busy state (held in the variable busy) of the micro:bit every 10 minutes and report the temperature every 20 seconds.

Setting the Variables

Our code uses the variables last_busy_flip and last_log_time to record the last time these two things (flip and log) happened:

```
busy = False
last_busy_flip = 0
busy_period = 600000
last_log_time = 0
log_period = 20000
```

The variable busy_period specifies the time in milliseconds between each flip between the busy and idle state; 600,000 milliseconds is equal to 600 seconds, which is 10 minutes. The variable log_period holds the time between temperature reports; 20,000 milliseconds is equal to 20 seconds. The code busy = False means the micro:bit starts off idle.

Making It Busy

If the micro:bit is in busy mode, both the display and radio are turned on and the message Busy is displayed. Otherwise, the radio and display are off.

```
while True:
    if busy:
     ❶ display.on()
     ❷ radio.on()
        display.show("Busy")
    else:
        display.off()
        radio.off()
    now = running_time()
    if now > last_busy_flip + busy_period:
        busy = not busy
        last_busy_flip = now
    now = running_time()
    if now > last_log_time + log_period:
        print((temperature(), busy * 10))
        last_log_time = now
```

In a while loop, we say that if the busy variable is True, the micro:bit should turn the display ❶ and radio ❷ on. Otherwise, they should be switched off.

Then we have two tests to see whether either a flip of the busy status or a log to the REPL and Plotter are due: the code checks the current runtime against the values in the busy_period and log_period variables.

To easily see when the processor is busy on the same plot as the temperature (Figure 9-1), we multiply the True or False Boolean value of busy by 10. Python allows us to do this! Rather than reporting an error, Python interprets False as 0 and True as 1. Because Python treats the Boolean values as numbers, it lets us multiply them. Then, instead of being either 0 or 1, the value is plotted as either 0 or 10.

How It Works: Why Does a Processor Heat Up?

A processor chip, like the one used by micro:bit, contains tens or even hundreds of thousands of *transistors*. These transistors are electronic switches that are in either an on or off state (represented by 1 and 0 in binary). Transistors use a very small amount of current when they are in a particular state (either on or off), but require a small amount of additional energy to change states. This is why, when doing something processor intensive, your computer fans will rev up—they're removing the excess heat generated by the large amount of switching.

A busy processor is a warmer processor. But other things can affect the processor temperature, too. In this experiment, the LEDs used by the display give off a little heat, as does switching the radio on and off. So, even though our definition of busy is a little misleading because it includes turning the radio and display on and off, the idea that the micro:bit's temperature readings are influenced by what the micro:bit happens to be doing still stands.

PROJECT: TEMPERATURE AND LIGHT LOGGER

Difficulty: Medium

As a hopeless perfectionist and incorrigible botanist, the Mad Scientist wanted a way to measure the ideal location for all the plants in the laboratory. And so a logger of temperature and light was born.

In Experiment 7, we wrote acceleration readings into a file for later analysis. In that case, we wanted to take readings immediately. Here, we want to log values over a period of time, say over the course of a day.

Figure 9-2 shows the completed project. As you can see, we have built it into a transparent food container to protect it from the elements, in case we want to use it outdoors.

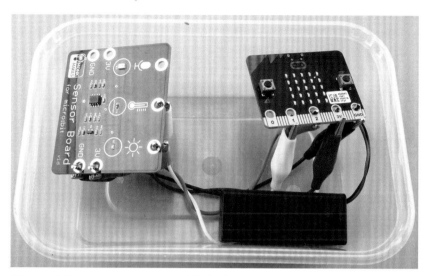

Figure 9-2: A temperature and light logger

You could use this logger to carry out a survey of your yard and determine the best place to grow different plants.

What You'll Need

For this project, you'll need the following items:

Micro:bit

Monk Makes Sensor for micro:bit   Another temperature and light sensor would work as well.

4 × Alligator clip cables

AAA battery pack

Transparent plastic food container   This is essential if you want to use the project outdoors. The container should be big enough to house the project, including the battery pack.

Construction

1. This project uses the micro:bit's local filesystem, which is not available in Blocks code yet. Therefore, this project can be done only in Python. Find the code at *https://github.com/simonmonk/mbms/*. The Python file for this experiment is *ch_09_Logger.py*. Flash the program onto your micro:bit.

2. Connect the micro:bit to the sensor board, as shown in Figure 9-3.

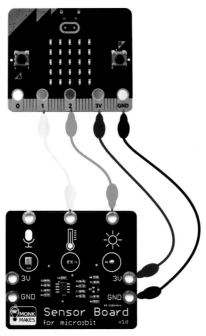

Figure 9-3: Connecting the micro:bit to the Monk Makes Sensor

3. Place everything into the food container, making sure that the sensor and micro:bit are at the top, near the lid. Fitting the alligator clips from the back of the boards will help.

4. Turn on the battery pack. When you're ready to start logging, press button **A** and then put the lid on the container. The display will change to show a single dot.

5. The logger can hold about 2,000 readings, so at a rate of one sample per second, it can run for 33 hours before running out of memory. This is about how long a set of AAA batteries should last.

When you're ready to stop the readings, press button **A**
again. You can then connect the micro:bit to your computer
and use Mu's *Files* feature to transfer the file *data.txt* onto
your computer, just as you did in Experiment 7 back in
Chapter 5.

6. To make sense of the data, you'll probably want to import
    it into a spreadsheet and draw some charts like the one in
    Figure 9-4. See Experiment 7 in Chapter 5 for an example
    of importing data from the *data.txt* file into Google Sheets.

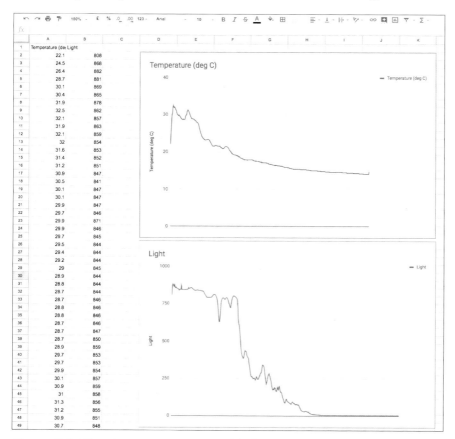

Figure 9-4: Charting temperature and light

In Figure 9-4, the steep drop of the line in the light chart
indicates the nightfall.

Code

For this project, we'll need to import the os library, which we'll use to delete any data file that might already be on the micro:bit. We do this to make space for our new readings. We'll also need log from the math library to write the code that calculates the temperature. *Log* (or *logarithm*) is a mathematical concept used in the Steinhart-Hart equation that converts the resistance measurement of a thermistor to a temperature. If you are into math, you might like to read more about logarithms here: *https://www .mathsisfun.com/algebra/logarithms.html*.

Setting the Variables

The code for this project waits for button A to be pressed and then repeatedly takes readings first of temperature then of the light level. Both readings are then written to a file so that when the logging is finished, they can be transferred a computer.

```
sample_period = 60000
filename = 'data.txt'
temp_pin = pin1
light_pin = pin2

last_sample_time = 0
recording = False
display.show(Image.NO)
```

We set the variable sample_period to 60000 milliseconds, or 1 minute. This period tells the micro:bit to take a recording once a minute. By default, the recorded data will be saved into a file called *data.txt*. You can change the name of this file by changing the value of the filename variable.

We tell the micro:bit which two pins are connected to the temperature and light outputs of the sensor board in the temp_pin and light_pin variables. In the last_sample_time variable, we store the last time a reading of the environment was recorded.

We use the variable recording to keep track of the whether the project is recording or not recording. This variable is toggled between True and False every time button A is pressed.

Reading the Temperature

We create the function read_c to read the temperature. This involves quite a lot of math.

```
def read_c():
    r0 = 100000.0
    r2 = 100000.0
    b = 4250.0
    v33 = 3.3 # actual result is independent of this value
    V = temp_pin.read_analog() * v33 / 1023.0
    R = r2 * (v33 - V) / V
    t0 = 273.15 # 0 deg C in K
    t25 = t0 + 25.0 # 25 deg C in K
    # Steinhart-Hart equation (google it)
    inv_T = 1/t25 + 1/b * log(R/r0)
    T = (1/inv_T - t0)
    return round(T, 1)
```

We won't go into the math in detail, but it's here in case you want to look through it.

This function measures the voltage at the temp_pin (pin 1) and uses that to calculate the temperature in degrees Celsius. See "How It Works: Sensors" on page 211 for more information on how this works.

If you want your temperatures in Fahrenheit, use the function read_f, which calls read_c and converts the temperature to Fahrenheit:

```
def read_f(self):
    return read_c() * 9/5 + 32
```

Reading the Light Level

The temperature readings are in units of Celsius or Fahrenheit. However, the light level is not expressed in any specific units. The light readings are just the direct analog readings from pin 2. In effect, we have defined our own units. You can give them a name if you like, perhaps *lightiness*?

Measuring light intensity in its standard unit of lux is difficult with this kind of sensor. But if you have a calibrated lux meter, you could carry out your own experiment to compare Lux and *lightiness* under different levels of illumination.

The while Loop

The main while loop (at the end of the code if you are following along in Mu) checks for a press of button A, toggling recording between True and False whenever the button is pressed. When recording starts, a single dot is displayed, and we delete the existing data file with os.remove. The remove command is contained within a try: except: Python structure. This makes sure that if an error occurs, probably because the data file isn't there and can't be deleted, the error is ignored and doesn't crash the program. After we remove the old file, the new file is opened with a mode of w for writing.

When button A is pressed again, the NO image is displayed and the file closed.

In the main while loop, there is also an if block that writes the readings from the light and temperature sensors to the file as long as recording is True and enough time has elapsed since the last_sample_time.

```
while True:
    if button_a.was_pressed():
        recording = not recording
        if recording:
            display.show(".")
            try:
                os.remove(filename)
            except:
                pass
            fs = open(filename, 'w')
        else:
            display.show(Image.NO)
            fs.close()
    now = running_time()
    if now > last_sample_time + sample_period:
        last_sample_time = now
        if recording:
            temp = read_c()
            light = light_pin.read_analog()
            fs.write(str(temp) + "," + str(light))
            fs.write('\n')
```

How It Works: Sensors

Thermistors are a special type of resistor (see the lie detector project in Chapter 8) whose resistance changes as the temperature changes. The type of thermistor used in the Monk Makes Sensor for micro:bit is an *NTC (negative temperature coefficient)*. The *negative* part means that when the temperature increases, the resistance decreases. We use the resistance to measure the temperature.

Our thermistor sensor measures temperature as electrical resistance. However, a micro:bit cannot measure resistance directly. Instead, resistance must first be converted into a voltage, and this can then be read by the micro:bit pin that's acting as an analog input. To do this, we need to use a voltage divider as we did in the lie detector project of Chapter 8. However, this time a thermistor, rather than someone's skin, will provide the variable resistance (Figure 9-5). Note that the Monk Makes Sensor for micro:bit board has the 100kΩ resistor built in.

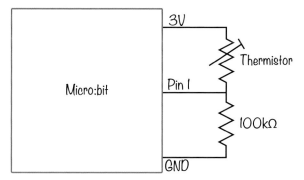

Figure 9-5: Schematic diagram for using a thermistor to measure temperature

A thermistor's resistance does not change every time there is a difference of some number of ohms, indicating a change of one degree. As you saw in the code earlier, the formula for calculating a thermistor's resistance is complicated, involving the use of logarithms.

When you buy a thermistor, it will specify two parameters:

▶ The resistance of the thermistor at 25 degrees C (called r0 in our code and equal to 100kΩ for the sensor board)

▶ A constant called beta, or sometimes just B, that is different for different thermistors (In our code, this is called b. For the thermistor on the sensor board, b is 4,250. The value of beta will always be specified on the datasheet for the thermistor.)

You can see how the calculation is made in the code. If you want to know more about this formula, search online for "Steinhart-Hart equation."

Because the voltage at pin 1 depends on the ratio of the resistance of the thermistor to the fixed resistor, it is independent of the supply voltage. This is just as well, because the 3V connector of the micro:bit can be anything from 3.3V down to about 2V, depending how you are powering the micro:bit and how fresh the batteries are. This is why you will see the comment in the read_c function explaining that the variable v33 (3.3V) has no effect. In fact, it cancels out in the math. It is included only to make the math a bit easier to follow.

PROJECT: AUTOMATIC PLANT WATERER

Difficulty: Hard

Ever busy with conferences that take them away from the Secret Lab, the Mad Scientist has devised this automatic plant waterer. The project monitors the resistance of the soil to determine how wet it is. If the soil gets too dry, it turns on a water pump. Pressing button A gives a readout of the soil dryness, and pressing button B acts as a test, running the pump for 10 seconds.

Figure 9-6: The automatic plant waterer project

What You'll Need

For this project, you'll need the following items. See the appendix for more information on where to find these.

Micro:bit

Relay board for micro:bit   To switch the pump on and off (You could also use a motor controller such as the Kitroniks board that we used in the rover project of Chapter 6.)

12V aquarium metering pump   These pumps are slow but reliable.

Tubing and connectors   To transfer the water from the reservoir to the plant pot

Large plastic bottle   To serve as a water reservoir for the plant

12V power supply for the pump   12V at 1 amp or more

USB power supply for the micro:bit or a Monk Makes Power for micro:bit and AC adapter This is a long-term project, so you don't really want to be running it from batteries. See the appendix for long-term micro:bit power options.

1kΩ resistor

Female DC barrel jack to screw terminal adapter To connect the relay and battery to the pump's power supply

7 × Alligator clip cables To connect the nails to the micro:bit

Fold-back binder clips To keep the watering tube in place

2 × 5-inch nails To be used as electrodes in the plant pot (Note that 6-inch nails also work just fine. Galvanized (zinc-coated) nails are best because they don't rust.)

Potted plant

The pump you need is a *peristaltic pump*, sometimes called a *dosing pump*. You can find one on eBay for a few dollars. These pumps usually come with short inlet and outlet tubes that you'll need to extend so they can reach from the water reservoir to the pot. They are generally 4 mm in diameter, a common size for garden irrigation systems. You can find such tubing and connectors at eBay, a hardware store, a tropical fish store, or a garden center.

You'll need longer alligator clip cables to connect the micro:bit to the nails. A good size would be about 1 foot (30 cm).

The water reservoir can be anything that will contain a reasonable volume of water (an old milk container would do fine).

Construction

You'll need to take extra care when building this project, as it has both an electronic and mechanical component. What's more, it pumps water around, so if you don't connect the pipes correctly, you could cause quite a mess in your Secret Lab!

Figure 9-7 shows how the electronics are connected to give you an idea of what you are aiming for as you follow the step-by-step instructions.

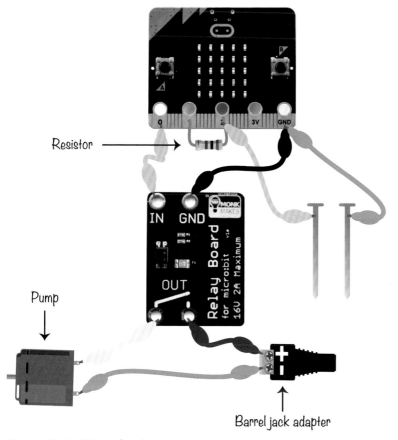

Figure 9-7: Wiring for the automatic plant waterer project

1. Open *https://github.com/simonmonk/mbms/* and click the link for **Plant Waterer**. Click **Download** and then copy the hex file onto your micro:bit. If you get stuck on this, head back to Chapter 1, where the process of transferring programs to your micro:bit is explained in full. The Python version of the code is in *ch_09_Plant_Waterer.py*.

2. Wrap the 1kΩ resistor's wires around pins 1 and 2 of the micro:bit.

3. Using Figure 9-7 as a reference, wire up the alligator clips to the micro:bit, relay board, barrel jack, and pump. Don't attach the yellow and green cables to the nails or connect the tubes to the pump just yet. To get the alligator clips to attach to the screw terminal ends of the DC barrel jack adapter, use a screwdriver to fully open the screw terminals. If your alligator clip jaws are too big to fit into the screw terminal holes, you can fit straightened-out paper clips into the screw terminals and then clip the alligator clips onto those.

4. Plug the 12V power adapter into the barrel jack adapter and then press button **B** on your micro:bit. You should hear the pump run for 10 seconds and then stop. If it doesn't do this, double-check your wiring.

5. Push the nails into the soil of the plant pots spaced about 3 or 4 inches (8 to 10 cm) apart. Leave enough of the nail above the soil that you can attach the green and yellow alligator clips, as shown in Figure 9-8.

Figure 9-8: Positioning the nails in the soil

6. Now, press button **A** on your micro:bit, and a number should scroll across the screen. This is a measure of the dryness of the soil. The drier the soil, the higher the number. Try adding a little water to the pot, wait for a few seconds, and then press button **A** again. You should see the number decrease. Don't make the soil too wet, though, as it will take ages to dry out again and you won't get to see the plant waterer in action.

   You now need to give the plant the right amount of water. You may need to consult someone with a green thumb who knows about your particular plant. Once you've found out how moist your plant's soil should be, keep adding small doses of water until the soil is damp enough. Then press button **A** and make a note of the number. This is your target dryness for the automatic plant waterer.

7. We're now ready for the wet part of this project. Start by measuring out suitable lengths of tubing from the pump. You want enough tubing that you can put one length through the top of the reservoir bottle and have it reach the bottom and the other will reach from the pump into the plant pot. Before attaching the tubing, you need to find out which nozzle of the pump is the *inlet* and which is the *outlet*. To do this, press button **B** and put a finger over each of the pump's tubes. You will feel the inlet pump sucking at your fingertip. Make a note of which tube is which.

   Use the tube connectors to attach these lengths of tubing to the pump's short tubes.

8. Clip the binder clips onto the side of the pot and push the tubing through the handles of the clip, as shown in Figure 9-9.

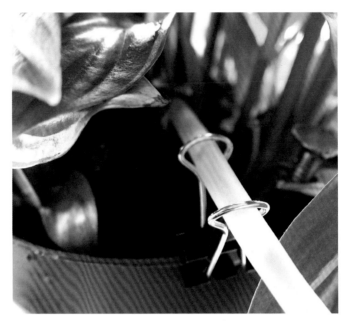

Figure 9-9: Securing the tube to the plant's pot

9. Fill up the reservoir bottle and push the extended inlet tube into the bottle, down to the bottom.

10. Test the pump by pressing button **B** a couple of times. If the water finds its way to the pot without any dripping or leakage, the waterer is almost ready to go. If not, find and seal the leaks by adjusting the connections.

11. The final step before you can let your plant waterer do its mundane work, leaving you free for more adventurous tasks, is to adjust the value of dry_threshold. Set dry_threshold to the value you recorded in step 6 and then flash the program onto your micro:bit again.

Code

The code for this project is quite complex. As well as monitoring the soil's moisture content, it also has to handle button presses and make sure that the pump doesn't get too carried away and flood the lab.

Blocks Code

Here is the Blocks code for the project.

```
on start
  set  dry_threshold ▼  to  500
  set  on_time_ms ▼  to  10000
  set  check_period_ms ▼  to  3600000
```

```
function  water_the_plant
  digital write pin  P0 ▼  to  1
  show arrow  South ▼
  pause (ms)  on_time_ms ▼
  digital write pin  P0 ▼  to  0
```

```
function  check_dryness
  digital write pin  P1 ▼  to  1
  set  dryness ▼  to  analog read pin  P2 ▼
  digital write pin  P1 ▼  to  0
```

```
forever
  if  running time (ms)  > ▼  dont_check_until ▼  then
    call check_dryness
    if  dryness ▼  > ▼  dry_threshold ▼  then
      call water_the_plant
    ⊕
    set  dont_check_until ▼  to  running time (ms)  + ▼  check_period_ms ▼
    plot bar graph of  dryness ▼
    up to  1023
  ⊕
```

```
on button  A ▼  pressed
  call check_dryness
  show number  dryness ▼
  plot bar graph of  dryness ▼
  up to  1023
```

```
on button  B ▼  pressed
  call water_the_plant
  call check_dryness
  plot bar graph of  dryness ▼
  up to  1023
```

In the on start block, we define three variables:

dry_threshold We put our value from step 6 here. If the plant gets dryer than this value, it will be watered.

on_time_ms This is the amount of time (in milliseconds) that the pump will run when watering. Keeping this value small (say 10 seconds) will keep the plant from being overwatered. It will also prevent accidents that might result in minor flooding!

check_period_ms The water needs a little time to spread throughout the pot and evenly wet the soil. This variable sets the delay between each dryness check. By default, it's set to 3,600,000 (1 hour in milliseconds).

If we have an on_time_ms value of 10, the maximum watering time the plant can receive in one day is 24 × 10 seconds or 4 minutes. With this kind of pump, it will receive about a pint (500 mL) of water. That's quite a lot, but if you have a really big pot, you may need to decrease check_period_ms or increase on_time_ms to allow the plant to get even more water. We'll discuss this further in "Things to Try" on page 222.

Besides these three variables, we have two functions, check_dryness and water_the_plant. The check_dryness function updates the dryness variable with a new soil reading from pin 2. Notice that this function also turns on pin 1, but just while the reading is being taken. We'll explain why in "How It Works: Measuring Soil Dampness" on page 222.

The water_the_plant function turns on pin 0 to activate the relay, turns on the pump for the time specified in on_time_ms, and displays the down arrow on the micro:bit to indicate that watering is in progress (a bit like it's raining).

With the forever loop, we first check whether sufficient time has elapsed since the last check (by default 1 hour). If enough time has passed, the loop calls check_dryness and compares this reading to the dry_threshold. If the pot is too dry, water_the_plant is called.

Now that the check is complete, the dont_water_until variable is set to the current time plus the check_period_ms to schedule the next check. The dryness is then shown on the display using the plot bar graph of block. The higher the level of LEDs on the display, the drier the soil is and the closer it is to being given more water.

Then we have the code that checks whether button A is pressed and reacts appropriately. This code calls check_dryness and then displays it before showing the bar graph again. The handler for button B calls water_the_plant and then displays the dryness level.

MicroPython Code

Here is the MicroPython version of the code:

```
from microbit import *

dryness = 0
dry_threshold = 500
on_time_ms = 10000
check_period_ms = 3600000
dont_check_until = 0

def water_the_plant():
    pin0.write_digital(1)
    display.show(Image.ARROW_S)
    sleep(on_time_ms)
    pin0.write_digital(0)

def check_dryness():
    global dryness
    pin1.write_digital(1)
    dryness = pin2.read_analog()
    pin1.write_digital(1)

def bargraph(a):
    display.clear()
    for y in range(0, 5):
        if a > y:
            for x in range(0, 5):
                display.set_pixel(x, 4-y, 9)

while True:
    if button_a.was_pressed():
        check_dryness()
        display.scroll(str(dryness))
        bargraph(dryness / 200)
```

```
    if button_b.was_pressed():
        water_the_plant()
        check_dryness()
        bargraph(dryness / 200)
    if running_time() > dont_check_until:
        check_dryness()
        if dryness > dry_threshold:
            water_the_plant()
        dont_check_until = running_time() + check_period_ms
        bargraph(dryness / 200)
```

Because MicroPython has no equivalent of the plot bar graph of block, we use the bargraph function from the Shout-o-meter in Chapter 2 to display the dryness level.

Things to Try

Because the plant waterer keeps the moisture level of the plant more or less constant, you can measure the amount of water the plant is using by seeing how much water has left the reservoir.

Use a measuring cup when you refill the reservoir bottle and log how much water you need to fill it back up. Once you know how much water your plant typically needs per day, you can work out how long your reservoir should last before it needs a refill. This will be very important if you want to keep your plant alive when you go on vacation.

How It Works: Measuring Soil Dampness

Impure water (such as water in soil) has a much lower electrical resistance than air. In other words, the dryer the soil, the higher its electrical resistance, and the more water in the soil, the lower its resistance. By measuring the resistance between the two nails, we can measure the dryness of the soil.

If you look back at Figure 9-7, you can see that there is a resistor between pins 1 and 2. To make it easier to see what is going on, another way of visualizing the schematic is shown in Figure 9-10.

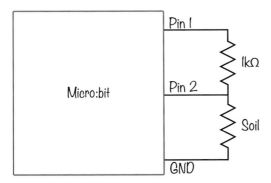

Figure 9-10: Schematic for measuring soil resistance

Notice that the diagram in Figure 9-10 is almost identical to the one in Figure 8-4 on page 198, where we were measuring skin resistance (rather than soil resistance) in the lie detector project. The one big difference is that rather than being connected permanently to 3V, the top of the fixed 1kΩ resistor is connected to pin 1. When we take a measurement of the soil dryness, we first set pin 1 high (to 3V) to take the reading and then set the pin back to 0V.

The reason for using Pin 1 instead of the 3V connection is that we want to allow electricity to flow through the soil only intermittently. If the resistor were attached to 3V, an electric current would always be flowing through the soil, messing up the readings and speeding up corrosion of the nails. This process is known as *electrolysis*. By only turning on pin 1 for the brief time that we take a reading, we avoid this problem.

SUMMARY

In this chapter, we explored how to measure temperature, created a temperature and light level data logger, and created an automatic plant waterer. In the next chapter, we'll see how the Mad Scientist uses the micro:bit's built-in radio library.

10

RADIO ACTIVITY

The Mad Scientist has made a friend. After hitting it off at a Mad Science Conference, the two decided they wanted to keep talking—through micro:bit, of course.

The micro:bit has a built-in radio transmitter and receiver, together known as a *transceiver*, that can communicate with Bluetooth devices. We saw this in action in the roving robot project in Chapter 6.

This radio transceiver can also be used to talk to other micro:bits using a simple message-sending and receiving protocol specific to the micro:bit. In this chapter, we'll look at micro:bit-to-micro:bit communication, so you'll need either two of your own micro:bits or a friend with a micro:bit.

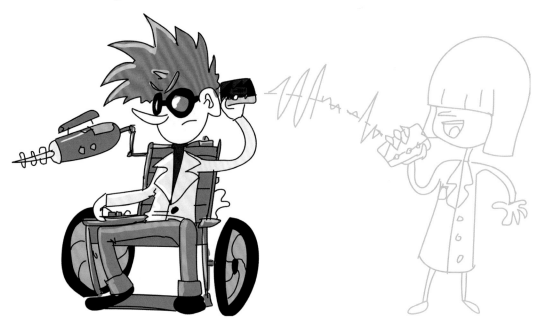

 # EXPERIMENT 12: FINDING THE RADIO RANGE

Difficulty: Easy

The Mad Scientist and a friend want to know how far apart they can be before their micro:bit communicators stop working.

What You'll Need

You'll need a pair of micro:bits, each equipped with a battery pack. You'll also need a friend to talk to and a field or other open space where you can move away from each other.

Construction

1. Go to *https://github.com/simonmonk/mbms/* and click the link for **Experiment 12: Radio Range**. Copy the hex file onto both micro:bits. (Chapter 1 explains the full process of getting programs onto your micro:bit if you get stuck). If you want to run the MicroPython version of this experiment, that file is *Experiment_12.py*.

2. Equip both micro:bits with battery power—you're probably used to doing this by now.

   Before you venture outdoors, test that both micro:bits are ready to go by pressing button **A** on one of the devices. An up arrow should appear on that micro:bit, and a check mark should appear on the other micro:bit. Repeat this test by pressing button **A** on the other micro:bit (Figure 10-1).

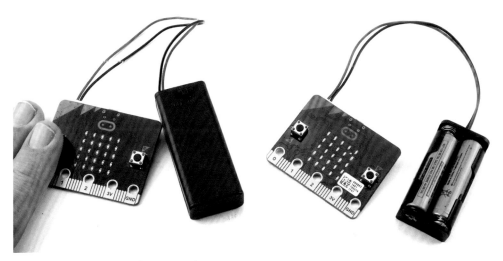

Figure 10-1: Testing the micro:bits

3. Go to a place where you and your friend have plenty of room to move away from each other. Stand about a yard (or meter) apart, facing one another. You should have one micro:bit and your friend the other.

4. Now either you or your friend presses button **A** on their micro:bit to transmit the signal. Then wave to let the other person know the signal was sent (in case they don't receive it). When their device picks up the signal, they should wave back. Assuming the signal was successfully transmitted, take a few steps away from each other and repeat the test.

5. At some point, the message won't be received! The sender should press the button and wave one more time. If the message still isn't received, you both know to move one step closer.

6. Once you've determined the radio's range, the sender uses a prearranged signal to tell the receiver to walk toward them, counting the number of steps that they take.

7. Measure the length of a step taken by the receiver and multiply that by the number of steps they took. The resulting value is the line-of-sight range of the micro:bit's radio. Tip: To get a reasonably accurate stride length, have the person walk five steps, use a long tape measure to measure the total distance traveled, and divide that distance by 5.

The Mad Scientist tested the range of two micro:bits and found it to be 192 strides. Five strides covered 12 feet (3.65 meters), meaning that each stride was 2.4 feet. The range was therefore 192×2.4 or approximately 460 feet (140 meters).

A range of over 1,100 feet (350 meters) has been reported by others carrying out this experiment. Note that the range will be considerably less if either your body or your friend's is between the micro:bits.

Code

The code for the range test is fairly simple, whether you use Blocks or MicroPython.

Blocks Code

Here is the Blocks code for the project.

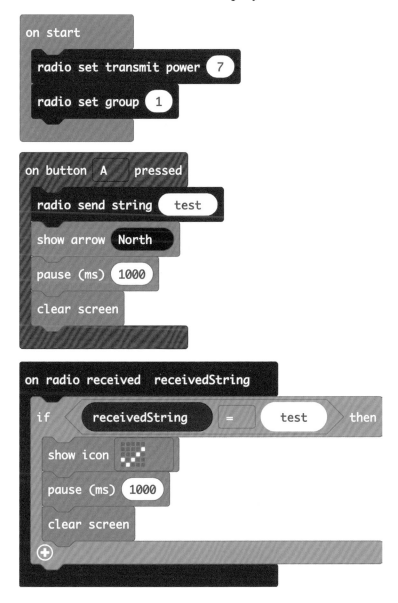

You can control the amount of power used by the radio. The on start block uses the radio set transmit power block to set the power to 7 (maximum)—more power means greater range. The set radio group block assigns a group for the radio to use. In this case, we use group 1, which means all the micro:bits set to radio group 1 will receive the transmissions. So, if you plan to carry out this experiment with multiple pairs of micro:bits at the same time, then each pair of experimenters should pick a different number between 0 and 255 and set the radio group to that number. This way, the different pairs won't interfere with each other.

We use the on button A pressed block with the radio send block to transmit a simple message of test. This will flash the North arrow icon to show that the message has been sent.

We handle incoming messages with an on received block, specifying the name of the variable into which any incoming message should be put. In this case, whenever the micro:bit's radio receives a message, it puts the message into the variable receivedString. If this message is test, then the check mark icon is displayed for a moment on the screen.

MicroPython Code

Here is the MicroPython version of the code. Note that the way messages are sent in MicroPython is slightly different from how they're sent in the Blocks code, so the pair of micro:bits used in this experiment should both be programmed either in Blocks or in MicroPython.

```
from microbit import *
import radio

radio.on()
radio.config(power=7, group=1)

while True:
    if button_a.was_pressed():
        radio.send("test")
        display.show(Image.ARROW_N)
        sleep(1000)
        display.clear()
    message = radio.receive()
    if message == 'test':
        display.show(Image.YES)
        sleep(1000)
        display.clear()
```

To set the radio's power and radio group in MicroPython, we need to use the `radio.config` method. This method also allows you to control a number of other options. You can read about these at *https://bbcmicrobitmicropython.readthedocs.io/en/latest/radio.html*.

The Blocks version of the code is contained entirely in handlers. In MicroPython, we don't have handlers, so we have to continually check for a button press or incoming message. We do this by having a set of `if` statements in a `while True` loop that is always running.

First, we check whether button A has been pressed since the previous check. If it has been, we send the message `test` and show the North arrow.

To check whether we've received a message, we repeatedly call `message.receive`. When the radio receives messages, it puts them in a queue. If there is no message waiting, then `message.receive` returns `None`. However, if there is one or more messages, then `message.receive` returns the oldest message and removes that message from the queue.

We only care whether the message is `test`, so we check this condition. If the message is `test`, we have the micro:bit display the YES icon for a second.

How It Works: Radio Signals

When outdoors, the *range* of the micro:bits' radios is likely to be much greater than when you're indoors, where walls between the micro:bits will impede the signal.

The system of sending messages between devices is called *packet radio* because small packets of data are being sent. In this project's code, these are text commands.

PROJECT: WIRELESS DOORBELL

Difficulty: Medium

Distracted by experiments and insubordinate underlings, the Mad Scientist often misses packages when they're delivered. To remedy this, they've decided to build a speaker in the lab that plays when the doorbell is rung.

We'll build on the doorbell project from way back in Chapter 2. In this version of the project, we'll use two micro:bits: one connected to a speaker, responsible for playing a tune, and a second one that acts as the doorbell button (Figure 10-2). When one of the buttons on the second micro:bit is pressed, it sends a radio message to the sound-making micro:bit, telling it to play a tune. Because the micro:bit radios have a pretty good range, your sound-playing micro:bit can be some distance from your door and, thus, closer to you.

Figure 10-2: The wireless doorbell project

What You'll Need

For this project, you'll need the following items:

2 × Micro:bit One that acts as a doorbell button and another that plays a tune

3 × Alligator clip cables To connect the micro:bit to the speaker

2 × USB power adapters or 3V battery packs with power switch To power the micro:bits

Speaker   To play the doorbell tune, I recommend the Monk Makes Speaker for micro:bit.

Blu-Tak adhesive putty or self-adhesive pads
To attach one of the micro:bits to the door frame

Construction

1. Go to *https://github.com/simonmonk/mbms/* and click the link for **Wireless Doorbell**. Copy the hex file onto both micro:bits. Chapter 1 explains the full process of getting programs onto your micro:bit if you need a refresher. If you want to run the MicroPython version of this experiment, that file is *ch_10_Wireless_Doorbell.py*.

2. Connect a speaker to one of the micro:bits. You can use the speaker you used in the musical doorbell project from Chapter 2 (see the instructions for this project if you get stuck).

3. Test the speaker by pressing button **A** on the doorbell micro:bit. The micro:bit attached to the speaker should immediately start playing the tune "The Entertainer." When it's finished, try pressing button **B**, and the "Funeral March" should play.

4. Use the Blu-Tak adhesive putty or pads to attach the speakerless micro:bit to the outside of your door.

Code

Both versions of the code rely on sending a message over the radio of either *db1* or *db2*, depending on which button is pressed. The receiving micro:bit then plays one of two tunes, depending on which message it receives.

As with Experiment 12, you cannot mix and match the MicroPython and Blocks versions of the code, so decide to use one or the other.

Blocks Code

Here is the Blocks code for the project.

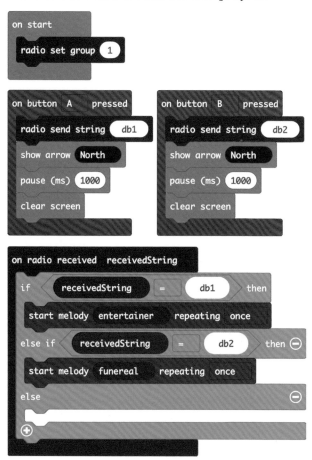

You'll notice that the code here is similar to that in Experiment 12. If Button A is pressed, then the string db1 (doorbell 1) is sent and a north arrow displayed to indicate that the message has been sent. The handler for when button B is pressed sends the message db2.

The receiving code checks whether the received message is db1 or db2 and plays the appropriate tune.

MicroPython Code

Here is MicroPython version of the code:

```
from microbit import *
import radio, music
```

```
radio.on()
radio.config(power=7, group=1)

def send_message(message):
    radio.send(message)
    display.show(Image.ARROW_N)
    sleep(1000)
    display.clear()

while True:
    if button_a.was_pressed():
        send_message("db1")
    if button_b.was_pressed():
        send_message("db2")
    message = radio.receive()
    if message == 'db1':
        music.play(music.ENTERTAINER)
    elif message == 'db2':
        music.play(music.FUNERAL)
```

In this version of the code, we've defined the function send_message, which sends a message using the radio and displays the north arrow for a second.

As with the code in Experiment 12, we use a while True loop to continuously check for button presses and received messages.

Things to Try

Try swapping in different tunes. Or you might try changing the code so that when db1 or db2 is received, the tune is played more than once. Then the Mad Scientist will be more likely to hear it!

How It Works: Sending and Receiving

You might be wondering why we use the same code for both the sender and the receiver. If we press button A on the micro:bit with a speaker attached, shouldn't it receive its own message and play a tune? It turns out that while the micro:bit's radio is busy transmitting, it cannot receive anything. Also, having just one program avoids confusion about which program goes on which micro:bit.

PROJECT: MICRO:BIT-CONTROLLED ROVER

Difficulty: Hard

No secret lab would be complete without a robot that can give instructions. Back in Chapter 6, we made a robot rover that could be controlled over Bluetooth using your phone. This project uses the same basic rover, but instead of controlling the rover with your phone and Bluetooth, you'll use a second micro:bit and the micro:bit's own way of communicating wirelessly. You'll steer the rover by tilting the controlling micro:bit left, right, forward, or backward. Figure 10-3 shows the project, and you can see it in action at *https://youtu.be/Qqr0fknoPQ4/*.

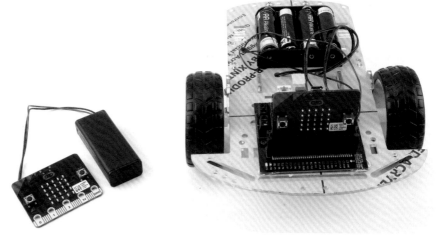

Figure 10-3: A micro:bit-controlled rover

What You'll Need

For this project, you'll need the following items:

2 × Micro:bit

Kitronik Motor Driver Board for micro:bit (V2)
To control the forward and backward motors

Low-cost robot chassis kit Includes two gear motors and a 4 × AA battery box

4 × AA batteries

AAA battery pack for micro:bit   To power the
micro:bit being used as the remote control

Screwdrivers   Suitable for both the nuts and bolts on
the chassis and the screw terminals on the motor controller board

Soldering equipment   To attach wires to the
gearmotors

Blu-Tack adhesive putty   To attach the motor control
board and micro:bit to the chassis

Construction

Use the rover you built in Chapter 6 or, if you haven't built
the rover yet, go back and follow construction steps 1 to 4
from that project. We'll be using different software, however,
so once the chassis is built, follow the instructions here. Don't

fit the batteries yet, or your rover might accidentally drive itself off your table!

1. First, we'll install the program for the rover part of the project. Go to *https://github.com/simonmonk/mbms/* and click the link for **Rover**. Copy the hex file onto the micro:bit attached to the rover chassis. Head back to Chapter 1 if you need more detailed instructions on how to get programs onto your micro:bit.

2. Now install the program for the micro:bit being used as the remote control. Go to the Github page, click the link for **Rover Controller**, click **Download**, and copy the hex file onto the micro:bit.

3. Before you let your robot loose in the lab, it's worth testing out your project without the wheels touching the ground. You may need to swap over some of the motor wires so the rover follows the commands you send it correctly. Insert the batteries and then flip the rover on its back so you can see what the wheels are doing, without any danger of its driving away.

   Tip the controlling micro:bit to the left, and you should see the same left arrow appear on the displays of both micro:bits. At the same time, both wheels should turn in the same direction. As you look from above, the right wheel should be turning faster than the left. If one of the wheels is turning in the wrong direction, swap over the red and black wires at the motor controller screw terminal for that motor.

4. Try driving the rover around. Remember that if the vehicle gets stuck, you can stop it by putting the controller micro:bit into a horizontal position.

Code

The software uses two programs, one for the controller and another for the rover micro:bit. You can only use Blocks code for this project

Controller Code

Here is the code for the remote controller micro:bit.

You'll notice that in the on start block, we use a radio set group block. This block makes the micro:bit listen only to messages from other micro:bits in the same group—here group 1. This prevents your micro:bit from picking up stray messages from other scientists in the area who might be using the same radio group and behaving unpredictably as a result. If you want to add other micro:bit pairs, change the number in the radio set group block to a different value for each micro:bit pair. Then each rover will be paired to a single controller. You can pick any number between 0 and 255.

The rest of the program consists of on gesture blocks that handle the possible movement commands for the rover. For example, below the on start block, you have the on tilt left block, which transmits the string L and displays an East arrow when the micro:bit is tilted. Here's the full list of commands that can be sent:

S Stop
L Left
R Right
B Backward
F Forward

Rover Code

Here is the receiving code for these commands.

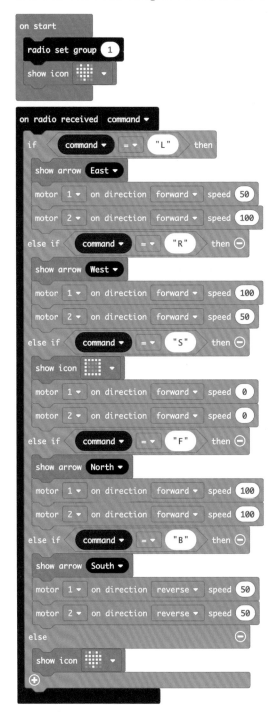

Similar to the controller code, we have an on start block that sets the radio group to 1. Remember that if you decide to change the radio group code, you have to do it on both micro:bits!

The rest of the code is contained in an on radio received block. Inside the block is a series of if statements that test the incoming command letter and perform the action that letter signals. If the L command is received, for example, a left arrow is shown, and then motor 1 is set to go forward at 50 percent and motor 2 to go forward at 100 percent (full speed). This will make motor 2 (the right motor) go faster than the left motor, making the rover turn in an arc toward the left.

Things to Try

Try adding extra commands to the pair of programs. You're getting a bit short on gestures to use, but you could add a D (for *dance*) command that tells the rover to do a little sequence of moves when the controller micro:bit is shaken.

You could also add a C (for *circle*) command that instructs the rover to spin on the spot by setting one motor forward at full speed and the other in reverse at full speed. This could be triggered by pressing button A or B.

How It Works: Motor Driver Blocks

You may have noticed a new category of blocks that appeared in your Blocks code when you opened the code for the rover: *Motor Driver blocks*. These blocks were created by Kitronik, the makers of the motor controller used in the project.

If you're starting a new project and want to use these blocks, you first need to add them to your project. To do this, click **Extensions** at the bottom of the list of block categories. This will open a dialog that looks something like Figure 10-4. If the package isn't listed after you search for it, refresh the browser page and try searching again.

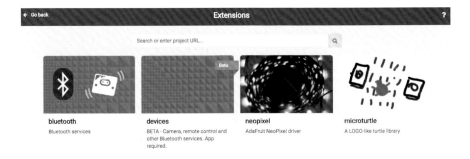

Figure 10-4: Managing extensions in the Blocks editor

In the field at the top labeled *Search or enter project URL . . .* enter the following: *https://github.com/KitronikLtd/pxt-kitronik-motor-driver/*. To make sure you get the URL right, enter it into another browser tab first. When you've found the page, copy and paste the URL from your browser's address bar to the field.

Once you've entered the URL, you should see *kitronik-motor-driver*, as shown in Figure 10-5. Click it, and you'll find that your Blocks editor now has a new category containing the motor control blocks that you can drag into your code.

Figure 10-5: Managing extensions in the Blocks editor

Once the package has been added to your project, it will be stored in the project forever. You won't need to install it again, unless you start a new project and want to use the package there. Because the package is stored in the project, you can easily share the project with someone else, with no need for them to install the package.

SUMMARY

In this chapter, you tested the range of the micro:bit's built-in radio, built a better doorbell, and made a remote-controlled rover. With its good range, the radio lets micro:bits communicate with each other easily, and it lends itself to all sorts of communication projects.

This is the final chapter in this book. The appendix that follows will give you some information about the parts you need to build the projects in this book and where you might obtain them.

The micro:bit community is a vibrant and active one. You'll find lots of interesting projects to make and experiments to carry out involving your micro:bit. Take a look at *https://microbit.org/ideas/* if you want some inspiration for what you, as a Mad Scientist, might do next with your micro:bit.

APPENDIX:
GET THE PARTS

he projects in this book use a lot of parts, and unfortunately, there is no one supplier that has all of them. If you find a project you want to build or an experiment you want to try, carefully look through the "What You'll Need" section and make a note of the things you don't have. This will be your shopping list.

The tables in this appendix will help you find the items on your list. If you're struggling to find something, do an internet search on the item's name or description, and you'll usually find somewhere you can buy it.

The Monk Makes Electronics Starter Kit for micro:bit (*https://monkmakes.com/mb_kit/*) will provide you with alligator clip cables as well as the Monk Makes Speaker and Sensor boards used in several of the projects in this book.

USEFUL TOOLS

Most of the projects and experiments in this book don't require any tools beyond your hands. However, there are a few tools that every Mad Scientist should have available. You probably already have many of these in your home.

Item	Description	Some Sources
Scissors		
Adhesive tape		
Screwdrivers	Flat head and Phillips (cross head)	
Drill (for Chapter 6)		
Soldering kit (for Chapter 6)	A low-cost kit is fine. You don't need anything fancy.	Ladyada's Electronics Toolkit: *https://www.adafruit.com/product/136*. For lower-cost alternatives, search Amazon and eBay for "soldering kit."
Craft knife	To cut shapes out of cardboard	Craft/hobby store
Pliers	Medium-sized general-purpose pliers are useful for bending wire.	Hardware store
Ruler (inches and cm)		Office supply store
Blu-Tack/Adhesive putty	This is great for sticking micro:bits to things in a nonpermanent manner.	Office supply store

Only the roving robot project in Chapter 6 requires any soldering. If you don't plan to make this project, you won't need a soldering kit.

Also in Chapter 6, the animatronic head project requires you to use a drill to make holes in ping-pong balls. Otherwise, you don't need this piece of equipment.

COMMON PARTS

Here are some of the parts that are used in many of the projects and experiments in this book. A number of micro:bit "getting started" bundles include some or all of these items, and some of these are listed here as well.

Item	Description	Some Sources
USB to micro USB data cable	To power and program your micro:bit	These cables are often used to charge cell phones, so you may have one lying around. If not, they are readily available from stores that carry cell phones or computers. They are also included in the micro:bit Go bundle and Pimoroni micro:bit accessories pack.
Alligator clip cables	To connect the micro:bit to various things	*https://www.adafruit.com/ product/1592 https://shop.pimoroni.com/ products/crocodile-leads-set -of-10/ https://www.kitronik.co.uk/ 2407-crocodile-leads-pack-of-10 .html*
Alligator clip to male header cables	To connect the micro:bit to a servomotor	*https://www.adafruit.com/ product/3255 https://thepihut.com/products/ adafruit-small-alligator-clip -to-male-jumper-wire-bundle-6 -pieces-ada3448/*

Here are some kits that contain some or all of these items.

Item	Description	Some Sources
Pimoroni micro:bit accessories kit	Includes a USB cable and battery box	*https://shop.pimoroni .com/products/micro-bit -accessories-kit/*
BBC micro:bit Go bundle	Includes a micro:bit, USB cable, and battery box	*https://www.adafruit.com/ product/3362*
Monk Makes Electronics Starter Kit for micro:bit	Includes alligator clip cables and a number of accessory boards	*https://www.eduporium .com/store/monk-makes -electronics-starter-kit-for -micro-bit.html* *https://thepihut.com/ products/electronics-starter -kit-for-micro-bit/*

POWERING YOUR MICRO:BIT

Here are some of the options available for powering your micro:bit.

Item	Description	Some Sources
2 × AAA battery pack	Holder for AAA batteries	Available in the micro:bit Go bundle: *https://www.adafruit.com/ microbit* *https://shop.pimoroni.com/ products/battery-holder-2-x-aaa -with-switch/* You can also find these on eBay.
Kitronik MI:power board	Coin-cell power for your micro:bit	*https://www.kitronik.co.uk/5610 -mipower-board-for-the-bbc -microbit.html*
Monk Makes Power for micro:bit	Powers the micro:bit from a 4.5V to 12V DC source, which is useful for long-term powering of the micro:bit	*https://www.kitronik.co.uk/46144 -monk-makes-power-board-for -microbit.html* *https://shop.pimoroni.com/ products/power-for-micro-bit/* *https://www.robotshop.com/en/ monk-makes-power-module -microbit.html*

Item	Description	Some Sources
ElecFreaks micro:bit Power Supply Module	The same concept as power for micro:bit (Note that the connecting cable is not included.)	*https://www.elecfreaks.com/estore/ micro-bit-power-supply-module-3 -3v-2a.html*
Charger kit for micro:bit	Rechargeable battery and case kit for micro:bit	*https://www.monkmakes.com/ mb_charger/*
USB backup battery	A useful rechargeable battery option (not suitable for high-current projects)	*https://www.adafruit.com/ product/1959* You can also find these at stores that carry cell phones or computers, as well as on eBay and Amazon.

MICRO:BIT ACCESSORIES

This book makes ample use of micro:bit accessories, like speakers and sensors. Here are some options for the accessories.

Item	Description	Some Sources
Monk Makes Speaker for micro:bit	Loudspeaker for sound-related projects	*https://www.eduporium.com/ store/monk-makes-speaker-for -micro-bit.html* *https://shop.pimoroni.com/ products/speaker-for-micro-bit/* *https://www.kitronik.co.uk/ 46124-powered-speaker-board -for-microbit.html*
Mini.Mu speaker	Loudspeaker for sound-related projects	*https://shop.pimoroni.com/ products/mini-mu-speaker/*

(continued)

Item	Description	Some Sources
Monk Makes Sensor for micro:bit	Sound, temperature, and light sensor	*https://www.eduporium.com/ store/monk-makes-sensor-for -micro-bit.html* *https://www.kitronik.co.uk/ 46122-sensor-board-for -microbit.html* *https://shop.pimoroni.com/ products/sensor-for-micro-bit/*
Adafruit MEMS Microphone Breakout	Advanced option for sound sensing (soldering required)	*https://www.adafruit.com/ product/2716*
Kitronik Motor Driver Board for the BBC micro:bit (V2)	For the robot rover project in Chapter 6	*https://www.kitronik.co.uk/5620 -motor-driver-board-for-the-bbc -microbit-v2.html*

MISCELLANEOUS

As well as the add-on accessories, you'll need a few other items.

Item	Description	Some Sources
Neodymium magnet (10 mm disc)	A very powerful disc magnet	You might find these in a hobby/craft shop, but eBay is probably your best bet. Search for "neodymium magnets."
3V Servomotor	Low-power servomotors that will operate at 3V	*https://www.adafruit.com/ product/169* *https://www.kitronik.co.uk/ 2565-180-mini-servo.html*
12V aquarium metering pump	Used in the plant waterer project in Chapter 9	Tropical fish store or eBay, searching for "12V aquarium metering pump"

Item	Description	Some Sources
1kΩ resistor	Used in the plant waterer project in Chapter 9	*https://www.sparkfun.com/ products/14492/* *https://www.kitronik.co.uk/ c3003-resistor-pack-of-100 .html* (version 3003-1k)
Female DC barrel jack to screw terminal adapter	Used in the plant waterer project in Chapter 9	*https://www.adafruit.com/ product/369* *https://shop.pimoroni.com/ products/male-dc-power -adapter-2-1mm-plug-to -screw-terminal-block/*
12V power supply for the pump	To power the pump in the plant waterer project in Chapter 9 (Pick one with your country's type of AC outlet plug that can supply 12V at 1A)	*US: https://www.adafruit .com/product/798* *UK: https://shop.pimoroni .com/products/power -supply-12v-1a/*

Micro:bit for Mad Scientists is set in Century Schoolbook, Filmotype Candy, Housearama Kingpin, and TheSansMono Condensed.

RESOURCES

Visit *https://www.nostarch.com/microbitformad/* for resources, errata, and more information.

MORE SMART BOOKS FOR CURIOUS KIDS!

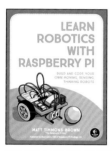

LEARN ROBOTICS WITH RASPBERRY PI
Build and Code Your Own Moving, Sensing, Thinking Robots
by MATT TIMMONS-BROWN
JANUARY 2019, 240 PP., $24.95
ISBN 978-1-59327-920-2
full color

CODING WITH MINECRAFT
Build Taller, Farm Faster, Mine Deeper, and Automate the Boring Stuff
by AL SWEIGART
MAY 2018, 256 PP., $29.95
ISBN 978-1-59327-853-3
full color

THE OFFICIAL SCRATCH CODING CARDS
Creative Coding Activities for Kids
by NATALIE RUSK *and the* SCRATCH TEAM
MAY 2019, 76 CARDS, $24.95
ISBN 978-1-59327-976-9
full color, box set

MAKE YOUR OWN SCRATCH GAMES!
by ANNA ANTHROPY
JULY 2019, 192 PP., $17.95
ISBN 978-1-59327-936-3
full color

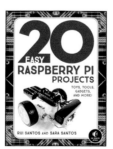

20 EASY RASPBERRY PI PROJECTS
Toys, Tools, Gadgets, and More!
by RUI SANTOS *and* SARA SANTOS
APRIL 2018, 288 PP., $24.95
ISBN 978-1-59327-843-4
full color

PYTHON FOR KIDS, 2ND EDITION
A Playful Introduction to Programming
by JASON R. BRIGGS
NOVEMBER 2022, 368 PP., $34.99
ISBN 978-1-7185-0302-1
full color

1.800.420.7240 or 1.415.863.9900 | sales@nostarch.com | www.nostarch.com